Wallentowitz · Amsel (Eds.)

42 V-PowerNets

Springer
Berlin
Heidelberg
New York
Hong Kong
London
Milan
Paris
Tokyo

H. Wallentowitz · C. Amsel (Eds.)

42 V - PowerNets

With 168 Figures and 16 Tables

Springer

Editors:

Prof. Dr.-Ing. Hennig Wallentowitz
Dipl.-Ing. Christian Amsel

RWTH Aachen
Institut für Kraftfahrwesen (ika)
Steinbachstr. 7
52074 Aachen, Germany

E-mail: wallentowitz@ika.rwth-aachen.de
amsel@ika.rwth-aachen.de

ISBN 3-540-44346-0 Springer-Verlag Berlin Heidelberg New York

Cataloging-in-Publication Data applied for
Bibliographic information published by Die Deutsche Bibliothek.
Die Deutsche Bibliothek lists this publication in the Deutsche Nationalbibliografie;
detailed bibliographic data is available in the Internet at <http://dnb.ddb.de>

Springer-Verlag Berlin Heidelberg New York
a member of BertelsmannSpringer Science+Business Media GmbH

http://www.springer.de

Typesetting: Camera ready by authors
Cover-design: Medio, Berlin
Printed on acid-free paper 62 / 3020 hu - 5 4 3 2 1 0 -

Preface

Preface to the Series

EUROMOTOR is an advanced training program sponsored by the European Union for engineers of the European automotive and supplier industry. Its aim is the exchange of experiences with new developments between industry and university. Courses are offered in all fields around the automobile, as for example vehicle dynamics, vibration & ergonomics, powertrain, crashworthiness, manufacturing, aerodynamics, testing and fatigue as well as courses in the field of body engineering and electronics.

For further information, please contact: EUROMOTOR Support Office, Caroline Marshall, RSBD, University of Birmingham, Birmingham B15 2TT, UK, Tel: +44 (0) 121 414 3695, Fax: +44 (0) 121 414 7151, e-mail: euromotor@bham.ac.uk.

Preface to this volume

On June 17th and 18th, the EUROMOTOR Course "1st Aachener Electronics Symposium" took place at the Institut für Kraftfahrwesen (ika) in Aachen/Germany. The conference dealt with the main topic

"Fusing Strategies for Future Vehicle PowerNets".

Speakers and participants from the automotive and supplier industry as well as research institutes reported the latest developments and trends in the field of *fusing strategies*. Technical experts in the field of active and passive fuses as well as systems suppliers presented special methods for the solution of the partial and overall problems. Intensive discussions about the technical aspects that have to be considered when introducing the new 42V-PowerNet into automobiles took place. The event was organized for a limited number of participants. For each presentation a time limit of 30 minutes with additional 15 minutes for a following discussion was set.

We hope that this compilation provides the reader with a well-balanced overview of the problems associated with guaranteeing the safety of future vehicle powernets. In order to ensure a continuous exchange of technology, this symposium is planned to be held every two years.

Our sincere thanks go out to the contributors for their eagerness to present the state-of-the-art technology from their organisations and for allowing their articles to be published in this volume.

Aachen, 2003

Prof. Dr.-Ing. Henning Wallentowitz
Dipl.-Ing. Christian Amsel

List of authors

Amsel, Christian
Forschungsgesellschaft Kraftfahrwesen mbH, Aachen, Germany

Bellew, Patrick
Littlefuse Ltd, Dundalk, Ireland

Borrego, Carles
Lear Corporation, Valls, Spain

Brinkmeyer; Horst
DaimlerChrysler, Sindelfingen, Germany

Edmund, Erich
Delphi, Wuppertal, Germany

Figuerola, Gabriel
Lear Corporation, Valls, Spain

Fischer, Roland
DaimlerChrysler, Frankfurt, Germany

Fontanilles, Joan
Lear Corporation, Valls, Spain

Graf, Volker
Intedis & Co. KG

Gresch, Peter
Adam Opel AG, ITDC-PE Electronics, Rüsselsheim, Germany

Gretzke, Werner
Tyco Electronics Raychem GmbH, Ottobrunn

Große, Ronald
Forschungsgesellschaft Kraftfahrwesen mbH, Aachen, Germany

Hinrichs, Werner
PUDENZ GmbH, 27243 Dünsen, Germany

Jasper, Jo
Litlefuse BV, Utrecht, Netherlands

Kroeker, Matthias
Tyco ElectronicsAMP GmbH, 13629 Berlin, Germany

Mäckel, Rainer,
DaimlerChrysler, Frankfurt, Germany

McLoughlin
Littlefuse Ltd, Dundalk, Ireland

Mestre, Jordi
Lear Corporation, Valls, Spain

Niebler, Erich
Vishay Electronic GmbH, Selb

O'Shea, Paddy
Vishay Intertechnology Inc., Ireland

Pohl, Klaus Dieter
Universität-Gesamthochschule Wuppertal, Germany

Seyer, Reinhard
DaimlerChrysler, Frankfurt, Germany

Scheele, Jürgen
PUDENZ GmbH, 27243 Dünsen, Germany

Schmidt, Fritz
DaimlerChrysler, Sindelfingen, Germany

Schulz, Thomas
DaimlerChrysler, Sindelfingen, Germany

Zörn, Ronald
PUDENZ GmbH, 27243 Dünsen, Germany

Contents

Introduction

The development of electronics in the automobile goes back 100 years. In the first seventy years, the development of electrical components was primarily restricted to elementary requirements such as ignition, lighting or windshield wiper control. The introduction of digital technology in the 70s revolutionized automotive electronics. Since then, constant development and improvements especially in the following fields have been achieved [MOS96]:

- safety,
- fuel consumption,
- exhaust emissions and
- comfort.

The development of additional sensors, actuators and control units resulted in a distinct increase in the demand on the performance of powernets. The first twenty years after the introduction of digital technology saw the mean power demand for upper-class vehicles double from around 0.6 kW to 1.2 kW. Rapid technical development today requires a power of 2.5 kW for an upper-class vehicle [SCT00].

Fig. 1. Conflict of objectives between weight, comfort and passive safety [WAL02]

One reason for the increased power requirement is the high demand on vehicle systems. Laws on fuel consumption and emissions, the further increasing traffic on roads, the desire for more comfort and safety as well as the pressure to reduce costs while simultaneously increasing the functionality, in any case lead to the increased use of powerful electronic regulators and control systems, hence resulting in a further increase in the power requirement of the powernet [WAL02].

The continuously growing power demands of the powernet could so far be met by increasing the generator power, battery capacity and cable cross-section. However the limitations of today's powernets are clear in middle and upper-class vehicles. Optional heavy-duty consumers such as electrical auxiliary heating or front windshield heating (each with a maximum power of approx. 1 kW) cannot be utilized up to the maximum extent with limited generator power due to their high power demand [GRE02]. An Energy Management System, which activates the heavy-duty consumers only in steps and dependent on the actual generator utilization, is required in order to maintain the availability of the power net [FRI01].

Fig. 2. shows the mean power demand of powernets with varying levels of auxiliary equipment in vehicles today as well as in the future. The electromagnetic valve train (EVT; average power approx. 1 kW) and the electrically driven air-conditioning compressor (average power approx. 2 kW), which power consumption takes place either over large intervals of time or continuously, are not considered here. Alternative approaches to distribute power to these systems using fuel cells to cover the basic load on the powernet have already been identified [PEH02].

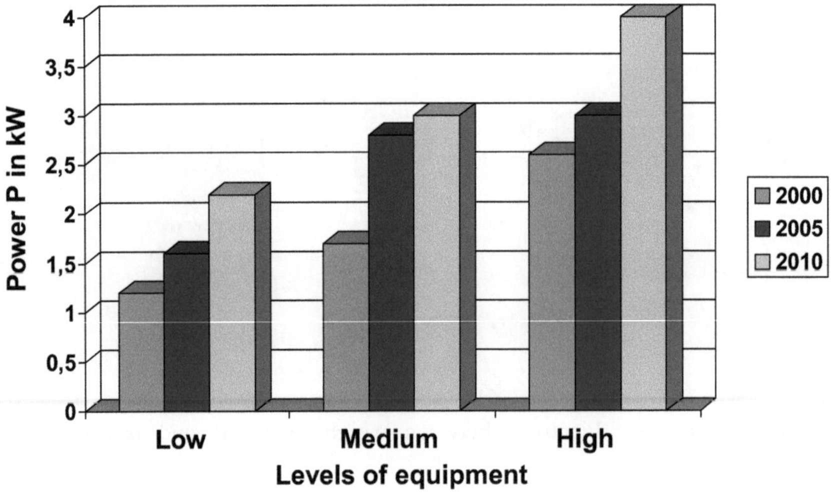

Fig. 2. Mean Power Demand in Powernets for varying levels of auxiliary equipment (without EVT and electrically driven air-conditioning compressor) [SCT00]

In order to measure up to the rising demands on the powernet for more generated power, the generator power has to be increased further, along with the optimisation of its efficiency. From today's point of view, generator powers of over 3 kW required in the future would no longer be sensible based on the technologies present at that time, due to constructional limitations and moments of inertia. The efficiency of these generators cannot be optimized beyond 70%. The introduction of Starter-Generators allows for the efficient generation of power above 3 kW.

The demands on the starter battery are always increasing, each time by a higher degree. On the one hand a battery with high values up to 110 Ah must meet short-term power demands of high values during cold starts. On the other hand heavy-duty consumers burden the battery in form of numereous charge/discharge cycles. An Energy Management System can reduce this effect only up to a limited extent. This has already led to the integration of two batteries in premium vehicles, each optimized in regard the demanded technology. These batteries are also expected to find application in vehicles of other segments.

The for future times expected rising pressure from regulations in regard to the reduction of fuel consumption and a further reduction in exhaust emissions [SCH02] leads to the introduction of starter-generator-systems with start-stop, boost and recuperative functions. The application of electrically assisted catalytic converters (peak power up to 3 kW) should lead to a further drop in the exhaust emissions during cold-starts.

The replacement of belt-driven auxiliaries by electrical systems which may be activated on demand (water pumps and servo-assisted steering), can lead to a temporarily increased consumer power. Still a long-term energy saving is possible, if the components are activated by demand. Improvements in comfort may be achieved by the broader application of electrical heating systems and new information and multimedia systems. Electrical locking systems, electrically operated steering devices (Electric Assisted Power Steering (EPAS), Active Front Steering (AFS)) and active chassis systems have an increasing demand on the performance, reliability and accessibility of the powernet.

An analysis and evaluation of today's situation and upcoming trends make it clear, that the present 14 V powernet (14 V represents the nominal voltage of the Generator; the nominal voltage of the starter battery amounts to 12 V) will no longer be able to meet the power demands in the mid-term future. The planned heavy-duty consumers would particularly require a higher powernet voltage for a technically and economically meaningful design [SCT00]. Infront of this background, already in 1998 it was decided that the powernet voltage would be increased to 42 V (42 V represents the nominal voltage of the Generator; the nominal voltage of the starter battery amounts to 36 V). So far however, the first serial vehicle with a limited functional range in 42 V is only available on the Japanese market.

A conversion of all systems in the vehicle from 14 V to 42 V will presumably not take place at once, hence creating the need to work with powernets with a combined 14 V and 42 V voltage level through the conversion phase.

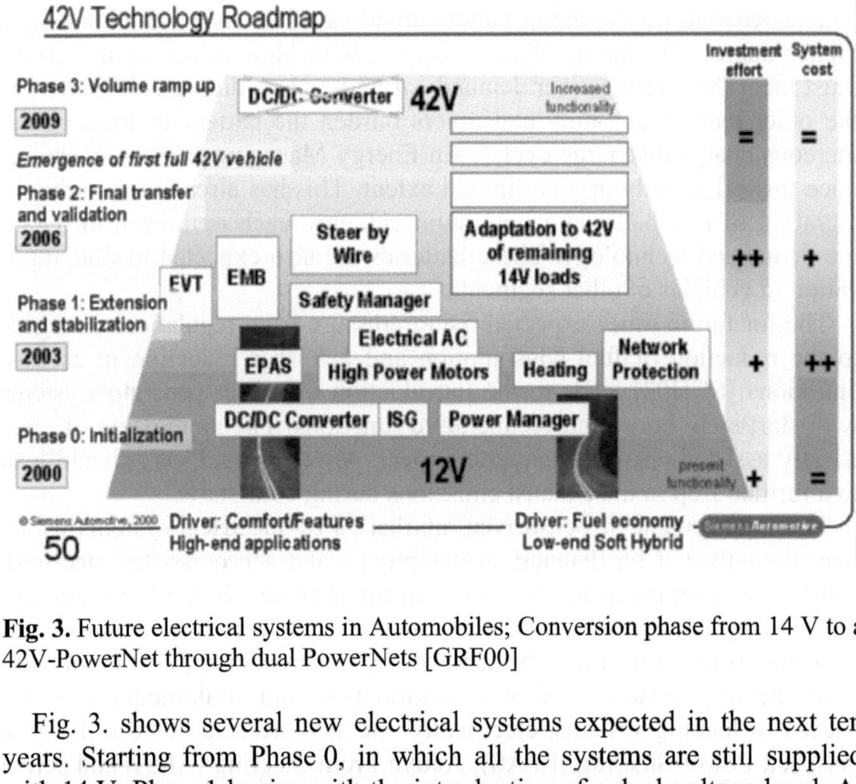

Fig. 3. Future electrical systems in Automobiles; Conversion phase from 14 V to a 42V-PowerNet through dual PowerNets [GRF00]

Fig. 3. shows several new electrical systems expected in the next ten years. Starting from Phase 0, in which all the systems are still supplied with 14 V, Phase 1 begins with the introduction of a dual voltage level. At the end of Phase 2, all systems are converted to 42 V and the first pure 42 V vehicles appear on the market. From Phase 3 on, all new vehicles would exclusively be supplied by a pure 42V-PowerNet.

The introduction of a 42V-PowerNet voltage would be primarily defined by the following factors:

- cost increase due to 14 V and 42 V variants,
- availability of 42 V components,
- system costs: costs for 42 V components (volume effects) and
- containment of all failures.

Guaranteeing safety in the vehicle is of always increasing importance due to the increasing complexity of powernets. The introduction of an additional voltage level of 42 V leads to new failure risks, some of which could have great consequences for the powernet. Short-circuits from the 42 V voltage level to the 14 V voltage level could damage substantial components of the 14 V powernet, as they are not designed for voltages in the range of 42 V. Another problem that arises is the increased risk of arcing at higher voltages. Arcing occurs e.g. when electrical contacts subjected to a load are withdrawn from their plugs (serial arc) and or in the case of damaged wiring (parallel arc). New concepts for protection must be developed and tested for both failure situations. In this volume, technical problems and their solutions associated with these topics are introduced and discussed.

This volume, containing the proceedings of the symposium, is structured into thirteen technical articles.

Dipl.-Ing. Peter Gresch, Hella Hueck KG investigates with his presentation „42V-PowerNet: Status of Evaluation, Requirements and Perspective“ the topic related to future powernets. After introducing the need and advantages of 42V-PowerNets, he illustrates various scenarios for the introduction of dual powernets. In his summary, Mr. Gresch emphasizes the dependence of the success of the 42V-PowerNet on a necessary standardization and the sufficient availability of components and systems.

Prof. Dr.-Ing. Dipl.-Chem. K. Pohl from the Bergische Universität-Gesamthochschule Wuppertal deals with the subject cable fires in his presentation „Cable Fire in Automobiles, Causes, Effects and Prevention“. Dr. Pohl introduces an experiment, which takes into consideration the influence of the resulting temperatures in automobile cabling depending on various cable cross-sections and insulating materials.

Following the introductory presentations, further technical presentations deal with different facets of the field of safety. Mr. Ir J. Jasper from Littlefuse B.V., Netherlands, talkes about solutions for voltage and current overloads in his presentation „Circuit Protection Components for Future Vehicle Electrical Systems“. Furthermore, he deals with fusible cut-outs (fuses) used in 42 V applications and current measurements with the help of shunts.

Dipl.-Ing. Dipl.-Wirt.-Ing. W. Gretzke of TYCO Electronics Raychem GmbH, Ottobrunn, considers the basic principles and applications of Polyswitch PPTC elements for protection against current and temperature overloads in his presentation „Use of PolySwitch PPTC Protection in Automotive Applications”.

Mr. O’Shea of Vishay Intertechnology Inc., Ireland, deals with solution possibilities for load dump, module level protection and ESD protection in his presentation on „Overvoltage Protection Devices for the Automotive Power Network".

Dipl.-Ing. W. Hinrich of Wilhelm PUDENZ GmbH, Dünsen, describes a new idea where the fuse is no longer designed on the basis of the consumer, but on the cabling. His presentation carries the name „New Fusing Concepts under Consideration of Wire Characteristics". In this way, he would like to prevent the over-dimensioning of cables.

Failures such as short circuits and arcs are the themes of the following presentations. Dr. M. Kroeker from TYCO Electronics AMP GmbH, Berlin, introduces new safety and protection stategies for arcing and short-circuits and presents methods for arc-detection and leakage current monitoring in his presentation "Power Switches for the 42V-PowerNet - Solution for PowerNet Protection and Applications".

Dipl.-Ing. R. Seyer of DaimlerChrysler AG, Frankfurt also deals with the phenomenon of arcing and presents detection and protection concepts in his presentation „Detection and Characterization of Short Circuits in 42V-PowerNets".

Dipl.-Ing. R. Große of fka Aachen, talks about a structured experimental method (Design of Experiments) for the analysis of failures in powernets in his presentation "Investigations of Electrical Faults in the Dual Voltage PowerNet". A special emphasis is given to the investigation of short circuits.

Holistic approaches to solve problems related to short circuits and arcing are introduced in the following three presentations. Mr. J. Fontanilles of LEAR Automotive (EEDS), Spain, presents an interesting approach for the protection of an electrical powernet without fuses in his article "Protection Strategies for Future Electrical Vehicle Architectures: Towards Fuseless Strategies". Intelligent semiconductors, which undertake both switching as well as protective functions, exclusively ensure safety in this concept. Dipl.-Ing. E. Erich of Delphi Automotive Systems, Deutschland GmbH, Wuppertal, introduces in his presentation „42V-PowerNet System Protection Concepts", methods to solve the problem of serial arcing at 42 V, which takes place when contacts subjected to a load are connected/disconnected. Furthermore, Mr. Erich reports about plugging concepts with corrosion protection.

In the presentation "Requirements for Introduction of the 42V-PowerNet" by Dipl.-Ing. V. Graf of Intedis GmbH & Co. KG, Würzburg, further analysis of failures, such as the loss of the common ground, as well as the formation of arcs are considered. Ground concepts as well as measures for the detection of arcs are discussed.

The presentation „Test Methods for Fusing Devices“ in which Dipl.-Ing. R. Zörn of Wilhelm PUDENZ GmbH, Dünsen, speaks about testing methods for fuse elements is the final contribution to the symposium.

Literature

[FRI01] Frielingsdorf B, Amsel C, Schoenen R, PTC Heizung, Haus der Technik, München, 2001

[GRE02] Gresch P, 42V-Bordnetz: Stand der Entwicklung, Anforderungen und Ausblick, Institut für Kraftfahrwesen Aachen, Aachen, 2002

[GRF00] Graf H, Einführung zukünftiger Energieversorgungssysteme – Neue Herausforderungen für die Systemintegration, Siemens Automotive, Regensburg, 2000

[MOS96] Moser O, Energieverbrauch 2002 – Verbraucher im elektrischen Bordnetz, Tagung Elektronik im Kraftfahrzeug, VDI, Baden-Baden, 1996

[PEH02] Pehr K, et al., Mit Wasserstoff in die Zukunft – der BMW 750hl, ATZ 2/2002 Jahrgang 104, Friedr. Vieweg Sohn Verlagsgesellschaft mbH, Wiesbaden, 2002

[SCH02] Schurk H, Kompaktseminar Kraftfahrzeugelektronik, Haus der Technik, München, 2002

[SCT00]Schöttle R, Threin G, Elektrisches Energiebordnetz: Gegenwart und Zukunft, Tagung Elektronik im Kraftfahrzeug, VDI, Baden-Baden, 2000

[WAL02] Wallentowitz H, Elektrische Fahrzeugsysteme , Umdruck zur Vorlesung Kraftfahrzeuge III, Aachen, 2002

42V-PowerNet: Status of Development, Requirements and Perspective

Peter Gresch

Adam Opel AG, ITDC-PE Electronics, Rüsselsheim, Germany

Abstract

The transition from a 12 V electrical vehicle system to a 42V-PowerNet is probably the most revolutionary change for electrical power distribution in decades. Many recent announcements in magazines, newspapers and various technical congresses and exhibitions forecast the introduction of 42 V components or even complete 42 V electrical systems in the near future.

The expectation is that nearly every automotive manufacturer will introduce a new vehicle using a 42V-PowerNet by the end of this decade.

1 Introduction

Especially on the European market, compact cars offer an electrical content, which was not even available in luxury cars some years ago. For example, the standard equipment of the Opel/Vauxhall Corsa in Western Europe includes two full size airbags, side airbags, seatbelt pretensioners and an electric power steering. Among the options are navigation systems or the GPS based telematics with OnStar service.

However, all these features are realized today, and they can be realized for the foreseeable future with 12 V – better referred to as 14 V power supply voltage - even though certain limitations have to be accepted.

For example, the Astra uses a 1 kW supplemental electric heater with the 2.0 Turbo Diesel for cold countries. A fairly complex algorithm is required to balance heating performance and battery discharge, gradually adjusting heating performance to the alternator output.

If gradual improvements lead to 42 V, it will not start in the compact class but most likely with the introduction of new features in upper class vehicles. These features cannot be offered with current 14 V (12 V) technology and may include:

- integrated starter/alternator,
- electrically heated windshield,
- electrically heated catalytic converter,
- electric waterpump,
- electric valvetrain,
- „x-by-wire", e.g. electrical steering,
- electro(-hydraulic/-mechanical) brakes and
- electric A/C.

Other features can be realized in 14 V technology but they face limitations and will greatly benefit from the 42V-PowerNet:

- electric engine cooling fan,
- electrical passenger compartment heating and
- active chassis control.

Forecasts estimate a power demand greater than 5 kW by the end of this decade:

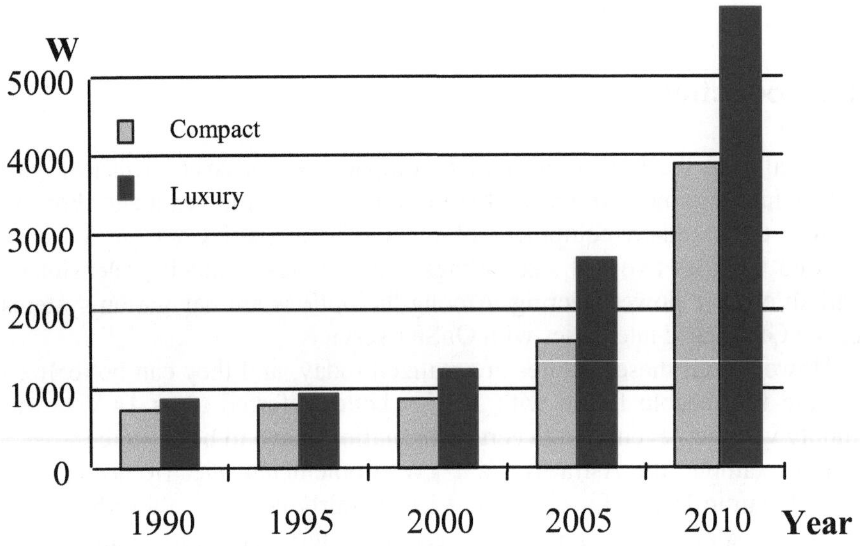

Fig. 1. Power consumption forecast

2 Reasons

The reasons for the transition from 14 (12) to 42 V can be divided into two major categories:

2.1 Gradual Improvement

The gradual increase of electric loads in vehicles leads to engineering challenges that ultimately enforce a higher voltage. Reasons are:

- alternators reach their performance limit, requiring complicated and expensive cooling strategies and
- wire gauges for power distribution become too big, adding excessive weight to the vehicle and causing manufacturing problems.

The transition to 42 V offers a whole set of new engineering solutions to improve existing systems which have reached their performance limit.

2.2 Enablers

Some electrical functions cannot be realized in the traditional 14 V technology. Hybrid systems with their demand for high electrical power for short periods of time fall into this category, as do x-by-wire applications. For these systems 42 V are a prerequisite.

Furthermore, the sum of all electrical loads in a new vehicle has to be taken into account in order to select the appropriate powernet configuration, namely 14 V or 14 V/42 V or 42 V.

3 Benefits

The main beneficiaries of higher voltages in vehicles are systems with high electric loads, for example electric/electro-hydraulic power steering or electric heaters. Integrating those on a traditional 14 V powernet causes problems that will not exist in the 42V-PowerNet.

Once the 42V-PowerNet is established, other loads will be added, replacing today's mechanically driven A/C compressors or water pumps with intelligent electric motors. Those will allow a flexible and continuous power adjustment unknown today.

Other advantages of the 42V-PowerNet include:

- higher efficiency of energy generation and distribution,
- improved controllability of electrically vs. mechanically powered consumers,
- weight reduction in the wiring harness,
- more efficient use of semiconductors for power switching and
- possibility for higher electrical loads.

At the International Technical Development Center in Ruesselsheim, Germany, Opel is working on a single voltage, single battery vehicle with local DC/DC conversion. For this research an Opel Astra has been modified to contain a 42 V starter/alternator, electro-hydraulic power steering, modified chassis control modules and a 36 V battery. The battery and a central body electronic module are located in the trunk, replacing the spare tire.

The basic structure of this vehicle's electrical architecture is shown in Fig. 2:

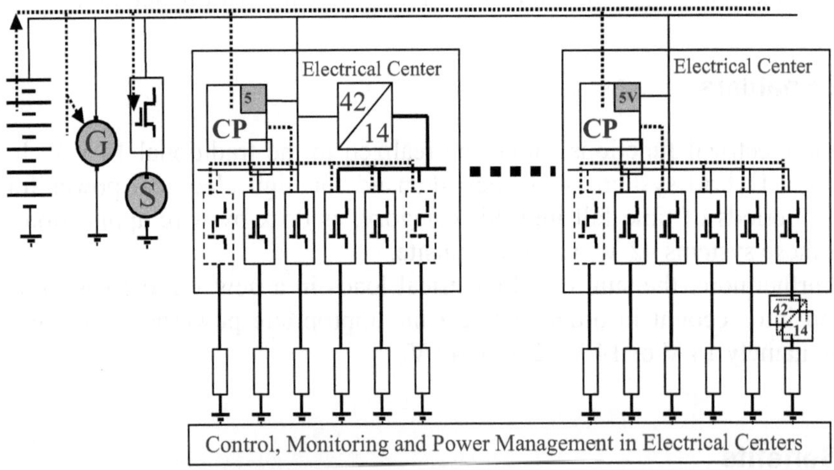

Fig. 2. Diagram of Opel 42V project

Any changes to the existing power generation and distribution system have to prove their superiority in respect to these requirements:

- improved functionality,
- weight reduction,
- quality improvement,
- package flexibility and
- improved fuel efficiency.

3.1 Fuel Efficiency

Often improved fuel efficiency is cited as a major driver for 42 V. Opel's recent analysis revealed that this assumption does not automatically hold true.

Fuel efficiency is determined in the MVEG (Motor Vehicle Emissions Group) cycle which is the basis of the ACEA (Association des Constructeurs Européens d'Automobiles) self-commitment to reduce CO_2 fleet emissions to 140 g CO_2/km (= 5.2 l/100 km gasoline) by 2008.

The average electrical load during the MVEG cycle is about 10 A. Generating those 10 A out of a traditional alternator with an efficiency of 50% causes a fuel consumption of approximately 0.13 l/100 km (see Fig. 3. and Table 1.).

The comparison of a 14V-FAS with predicted 75% efficiency with a 42 V FAS of about 80% efficiency results in an additional fuel efficiency improvement by 0.006 l/100 km. This efficiency gain is equivalent to a weight reduction of only approx. 4 kg.

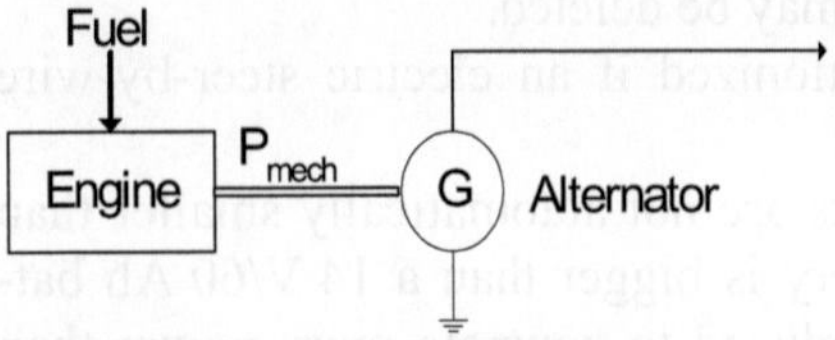

Fig. 3. Fuel Efficiency Calculations

Table 1. Data overview for different alternators

Alternator	Efficiency [%]	P_{mech} [W]	Fuel Consump. [MVEG]	Mass Equivalent
KCB2-14 V-100 A	50	280	0,130 l/100 km	base
E 120 A, improved efficiency	61	230	0,107 l/100 km	- 0,023 l/100 km ~ 15,3 kg
FAS, 14 V	75	187	0,087 l/100 km	- 0,043 l/100 km ~ 28,7 kg
FAS, 42 V	80	175	0,081 l/100 km	- 0,049 l/100 km ~ 32,7 kg

If 42 V can be related to fuel efficiency improvements, it is by enabling other features e.g. electric valvetrains or electrically heated catalytic converters.

3.2 Weight Reduction

The expectation that 42 V will lead to significantly reduced weight in wiring harnesses is only partially correct. An analysis based on existing harnesses has shown that the wiring of an Opel Astra would be about 4 kg lighter if 42 V components were used. The majority of wires today are signal wires; their gauge of 0.35 mm^2 is constrained by their mechanical strength and not affected by 42 V. In a dual voltage 14/42 V vehicle the weight improvement gained in the wiring harness would easily be overcompensated by the additional battery and DC to DC converter supporting the dual voltage architecture.

3.3 Package Flexibility

Among potential other benefits of 42V electrical systems is increased package flexibility. If the alternator is replaced by a flywheel alternator/starter and the air-conditioning (A/C) compressor is powered with an electric motor, the engine's belt drive may be deleted.

The interior design can be revolutionized if an electric steer-by-wire sensor replaces the steering column.

On the other hand, 42V components are not automatically smaller than 14V components. A 42 V/20 Ah battery is bigger than a 14 V/60 Ah battery. 42 V electrical systems are introduced to generate more power than today's 14 V systems. In general, this will lead to an increase in the overall size of electric power generation components rather than a decrease.

3.4 Performance Improvements

The usage of an electrically driven compressor will improve heating, ventilation, and air-conditioning (HVAC) performance at idle by utilizing full compressor capacity under this condition. This is currently impossible with belt driven systems. The compressor design may also become simpler with no need for variable displacement compressors.

Brake performance may be increased by using electric brakes and also save space where the traditional brake booster or hydraulic lines were once located.

Even though 42 V enable some functionality and solve some problems related to 14 V systems, there are still inherent challenges to it. Thus we will still see innovation in 14 V components. In contrast to earlier assumptions a Flywheel Alternator Starter for small engines can, for example, can be realized in 14 V technology.

4 Challenges

Undoubtedly the 42V-PowerNet offers many benefits; yet it also creates fundamental challenges. Many of these challenges related to a higher bus voltage of 42 V are still not entirely solved. Areas of particular interest when optimizing the power generation and distribution system are:

- electro-magnetic compatibility (EMC),
- quiescent current,
- protection against short circuits between 42 V and 14 V,
- connection systems, corrosion, sealing of wires and pins, ...,
- jump-start,
- power supply, grounding concepts,
- packaging, especially in small vehicles and
- cost.

4.1 Electro-Magnetic Compatibility

The 42V-PowerNet requires the increased usage of switching power supplies in contrast to linear regulators most commonly used today. These require additional efforts to meet current EMC standards, as well as additional parts and filters to reduce the radiated emissions from those converters, increasing overall module cost.

The first EMC tests conducted with a 42 V concept vehicle at Opel showed that radiated emissions would be the most significant difference between 14 V and 42 V components.

4.2 Quiescent Current

Also, there is no clear solution on how to handle the vehicle's quiescent current. If the energy stored in the vehicle's battery is the same in a 42 V car as it is today, consequently its Amp*Hour rating is reduced to one third of its 14 V value.

Hence, the typical 500 µA (at 14 V) quiescent current for an electronic module has to be reduced to one third of this value, or 166 µA (at 42 V). This means the impedance of the module in the "key-off" state has to increase by a factor of 9!

It is unlikely that this can be easily achieved without an architectural impact. Even worse is the substitution of relays for transistor switches in the 42 V architecture. Ironically, components with zero quiescent current are replaced with components which cause parasitic current in the architecture that can afford it least.

Providing very small currents (about 200 µA) is another challenge. Step down converters cannot be used continually due to their own quiescent current. A small linear regulator dedicated to support the processor sleep mode on the other hand would have increased losses in comparison with its 14 V alternative. It seems that partially active step down converters offer a technical solution while at the same time creating a new challenge for EMC.

Nevertheless, the Opel demo vehicle shows potentials to reach about 15 mA quiescent current in the long run.

Fig. 4. Test Bench „Opel 42V-Project"

4.3 Arcing

Arcing is another technical challenge that is unique to higher (42 V) voltage level. An arc can be established and maintained at voltages greater than about 20 V. Arc temperatures can reach 10,000°C, deforming surrounding plastic or causing even worse damage. Fuses are not an adequate protection against arcing, since the current of the arc does not need to be very high.

Thus, new solutions must be developed in a variety of areas, such as replacing fuses under load or dealing with the arcing caused in brush-type DC motors. Among other potential solutions is the usage of (Power)MOSFETs in the 42V-PowerNet instead of relays.

4.4 Power Supply

If a 42 V car is built with a conventional lead acid battery, several new challenges arise. By tripling the number of cells in the battery the potential for failure is also tripled. Using features like start/stop vehicle operation results in a different battery usage profile, increasing the number and discharge depth of battery cycles, thus causing additional stress to the battery life.

Assuming the battery failed, roadside assistance also poses a new challenge: Jump-start of a 42 V vehicle with a 14 V vehicle.

As a result, it becomes obvious that the energy storage in 42 V vehicles will be more sophisticated than today. Ultra-Capacitors (Ultracaps) could provide an alternative for short-period, high peak load storage of electrical power; yet, they are currently too high cost.

4.5 Optimization of Power Generation System

The 42V-PowerNet is introduced to support high current/high performance consumers. Those have very different characteristics. While an electric A/C and electric valvetrains are continuous high power consumers, which create challenges for the entire electric power generation, x-by-wire applications are most challenging because of very high peak currents. For these applications higher voltages seem necessary to limit the voltage drop in supply wires.

The Power Generation System has to accommodate both types of electrical loads, so it must be optimized for diverging requirements.

4.6 Cost

The compact and midsize car markets in Europe and America are extremely cost sensitive. Customers are willing to pay for functions; the technical realization is of minor importance. The marketing value of 42 V is marginal compared to its cost of introduction.

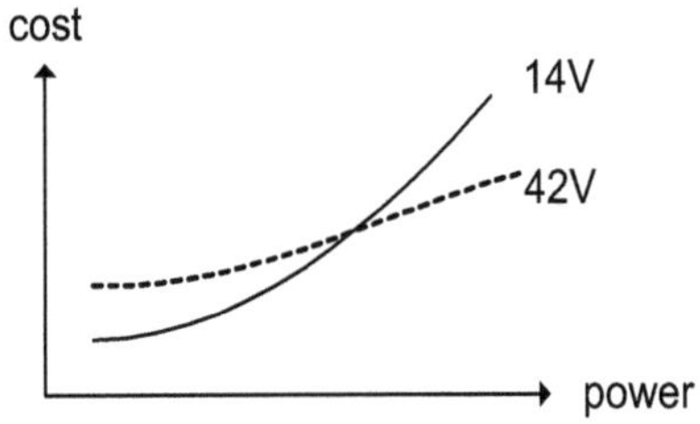

Fig. 5. 42V-PowerNet may become cost efficient if power demand increases

It is very unlikely that any OEM will go directly from a 14 V to a complete 42 V vehicle. Some elements like the voltage used at the diagnostic connector are mandated by law and are beyond the manufacturer's control. The architectural mix of 14 and 42 V components always requires additional parts (like a DC/DC converter or even a second battery) between the voltage levels, adding structural cost to a vehicle.

Though an additional cost burden of the dual voltage cannot be avoided, the system is not necessarily more expensive than a "pure" 42 V system. 42 V components will require new production lines, new development, and they will initially have a relatively low volume. Basic economic considerations indicate that early 42 V components will be more expensive than 14 V components to cover this additional fixed cost.

Even after years of mass production the variable cost of 42 V components may still exceed that of 14 V components because of more difficult EMC filtering or higher process cost. For example, the number of windings in a 42 V electric motor is three times more than in a 14 V motor, resulting in longer manufacturing time.

Given the competitive environment of the automotive industry worldwide, additional cost is unacceptable unless the value for the end customer outweighs the cost penalty.

5 Introduction Scenario

The development of 42 V vehicle systems offers unique opportunities. Obviously the number of independent car manufacturers is decreasing. Those who are part of a bigger corporation, such as Opel, have to find ways to intelligently use the economies of scale a bigger corporation offers.

Under the competitive goal of shortening the development time of new vehicles, common electrical parts or families of subsystems must be shared among vehicle platforms. This offers an opportunity to design vehicles faster and more cost efficient. By adding volume to parts that are used by different platforms it is possible to have more variations of partitioning than before, thereby tailoring a vehicle to an individual buyer's demands more than ever.

A new electrical architecture, such as the 42V-PowerNet, offers the "clean sheet" approach. Because there are no previous parts, there is a chance to develop a complete electrical architecture based on common components. General Motors will pursue this goal, as standardization is a key strategy for success. This is true both inside GM, its subsidiaries and allies as well as among all car manufacturers.

In those areas that are not competitive for the OEMs, standardization of 42 V interfaces should be a goal. There is no reason why a high side driver that is capable of driving one Amp must be different between vehicle manufacturers.

Furthermore, the automotive market compared to the consumer market is for semiconductor manufacturers small. Further division of this market into semiconductors that meet different OEM specifications forces additional development and validation efforts adding a cost penalty the vehicle customer has to pay.

Standardization of hardware interfaces among OEMs and suppliers should be a strategic goal that can be reached easier with the joint development of a new electrical architecture in automobiles: The 42V-PowerNet.

General Motors announced its 42 V strategy by introducing a portfolio of hybrid vehicles ranging from cars to full-size trucks in 2001 at the North American International Auto Show. The ParadiGM* system (say PAIR-a-dime) combines a V-6 or an inline-4 cylinder engine (ICE) with a pair of electric motors and a battery pack. The 3.6-liter V-6 version of the hybrid powertrain, which will be offered first, puts out 220 horsepower from the ICE, plus another 32 hp from the 42 V electric motors.

In addition, with a stronger focus on potential new vehicle features for the European market, Saab and Fiat recently finished a 42 V architecture study for "Premium Vehicles" (Fig. 6.) with the following main architecture design:

- the PowerNet architecture is a dual voltage architecture,
- the power is generated in the 42 V primary net,
- one DC/DC converter supplies the 14 V secondary net with power and
- one battery on each net (Gen III) to handle peak power and key-off current.

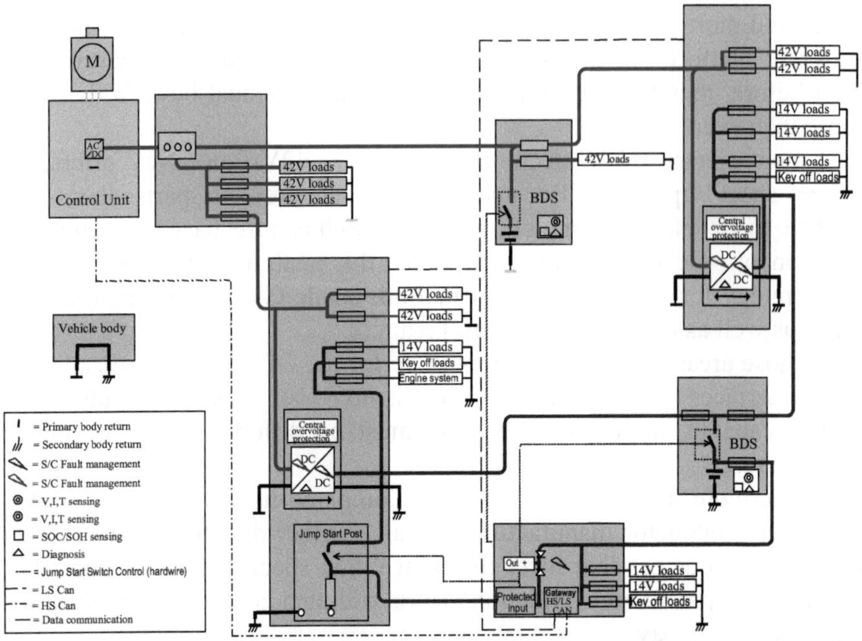

Fig. 6. Proposal for „Premium Vehicle" Architecture

6 Conclusion

In the long run cars will be more electric than today, converting traditional mechanical loads into more powerful and intelligent electric actuators. The 42 V electrical architecture will thereby act as the enabler of new technology in cars.

Yet, for small cars 14 V systems will stay very likely, at least for this decade, superior if not even for much longer than that.

The dual voltage, dual battery 14/42 V electrical system comes at a high structural cost and at least 16 kg additional weight. This system will only be used in a transitional period before pure 42 V systems can be introduced. The transition time should be kept as short as possible.

The comparison of the traditional 14 V system with a dual voltage 14/42 V system and a single voltage 42 V, however, also shows significant difficulties related to 42 V that still have to be solved (see Table 2.).

Table 2. Advantages and disadvantages of different power distribution systems

14 V		14/42 V		42 V	
+	Proven technology	+	Prepared for future electrical loads	+	Prepared for future electrical loads
+	Optimized cost	+	Starter/alternator without electrical limitation	+	Starter/alternator without electrical limitation
				+	Reduced weight of wiring harness (app. 4 kg)
-	Load management required	-	No jump start	-	No jump start
-	Starter/alternator only with small engines	-	Short circuit 42 V -> 14 V	-	Reduced standing time
-	Existing electrical loads can be supported, but limited reserves	-	New technology, little experience, few components available	-	New technology, little experience, few components available
		-	Package-, cost- (min. 100 Euro) and Weight penalty (16 kg)	-	High development risk
				-	Higher cost
				-	Transition possible only with major facelift, introduction not possible in next years

Other areas that are affected by the 42V-PowerNet are:

- perceived safety,
- corrosion,
- support of after market loads and trailers,
- change of manufacturing plants to handle 42 V and/or 14 V vehicles,
- training of service personnel and
- updating or new development of test and validation tools.

These technical challenges regarding the transition from 14 V to 42 V will influence the entire automotive value chain: From development through production to service.

As mentioned before, the development time when an architectural study is initiated until the start of production is continually decreasing, leaving less time for several development loops of electronic modules. As a result, volume production of 42 V vehicles can only happen if all sub-components like transistors, CAN transceivers, etc. are already developed and validated.

Finally, the 42V-PowerNet will only be successful if automotive manufacturers and suppliers and their development partners and sub-suppliers will follow the path of standardization.

7 References

[1] Gehardt K; Gresch P; Nix A; DR. Heitkämper P, Das 42 V Bordnetz aus der Sicht eines Volumenherstellers, VDI-Tagung "Die Architektur des 42 V Bordnetzes", Munich, October 13, 2000

[2] Gresch P; Nix A, The Future of 42 V Vehicle Systems, Second Ricardo International Conference, Brighton, June 13, 2001

[3] Gresch P, Anforderungen an das 42 V Bordnetz, ZVEI Veranstaltung 42 V Bordnetz, Frankfurt, October, 2001

Cable Fire in Automobiles: Causes, Effects and Prevention

Klaus Dieter Pohl

öffentl. Best. u. vereidigter Sachverständiger
Universität-Gesamthochschule Wuppertal
Fachbereich 14, Sicherheitstechnik

1 Introduction

Fires (Fig. 1.) or explosions are more less rapidly taking place exothermic oxidation reactions which when uncontrolled and self-sustaining, can lead to damage. An essential condition for fire is the appearance of a flash, heated or glowing surfaces (e.g. overheated metals such as boiler plates or metallic filaments in electrical lines) however being excluded.

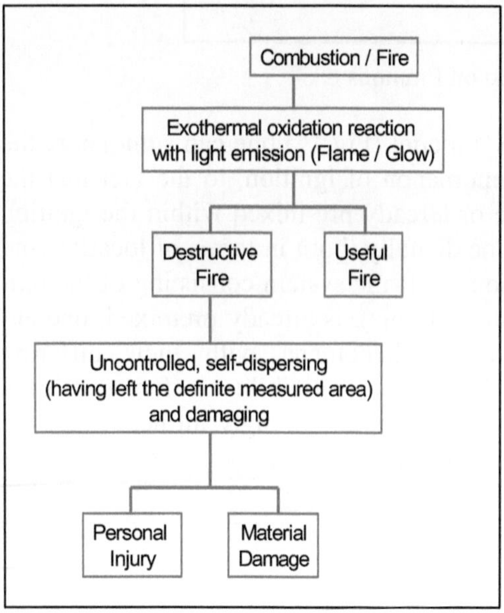

Fig. 1. Definition of fire according to insurance laws

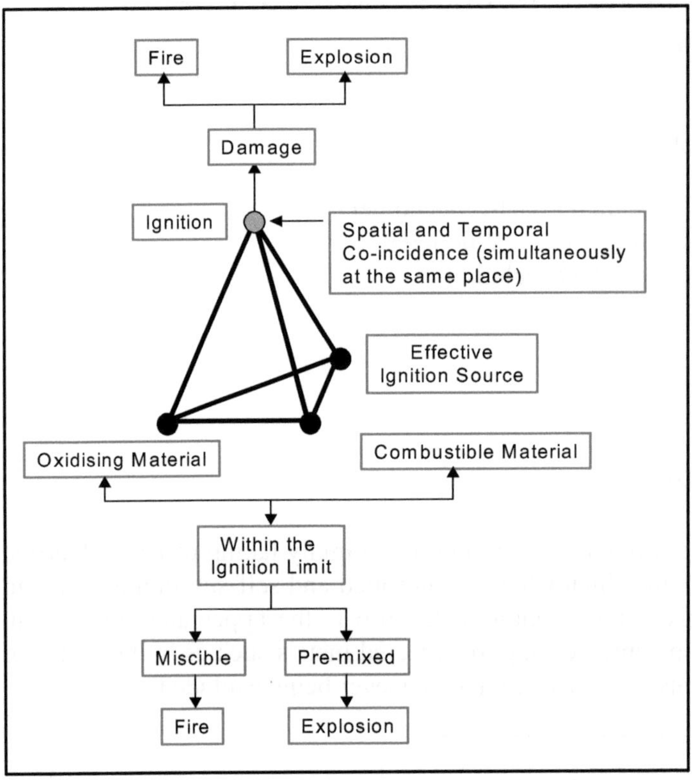

Fig. 2. The ignition tetrahedron based on Emmons

On the basis of the EMMONS's model (Fig. 2.) one can summarize the conditions which lead to the phenomenon of ignition, to the fact that the fuel and the oxidant are miscible or already pre-mixed within the ignition limits, and come into contact coincidentally, both in terms of location and time, with an effective ignition source. If the system consisting of the oxidant and the fuel (within the ignition limits), is already premixed, one enters into the group of explosion phenomena with their differing characteristics of damage (Fig. 3.). If only miscibility is foreseen, only fires result, i.e. Fire phenomena are in addition strongly influenced by the type of fuel (Fig. 4.) available with respect to the velocity of the reaction and characteristic of damage, without having to go into further differentiation here.

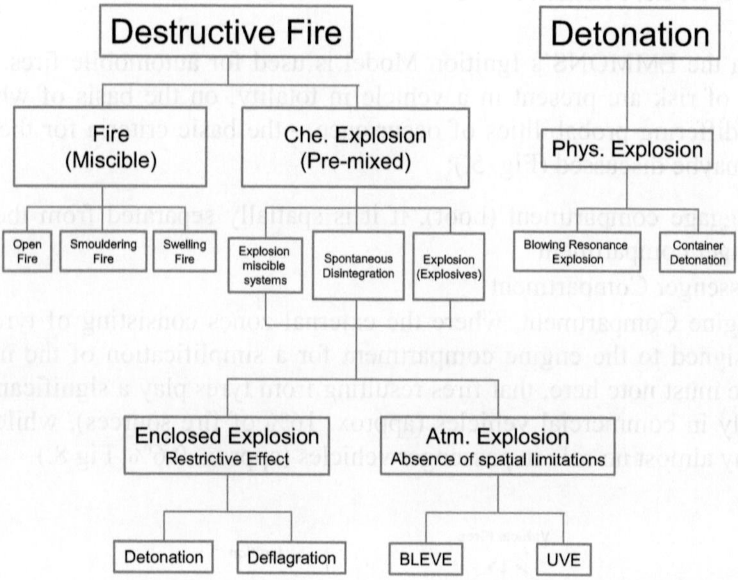

Fig. 3. Characteristic of Damage

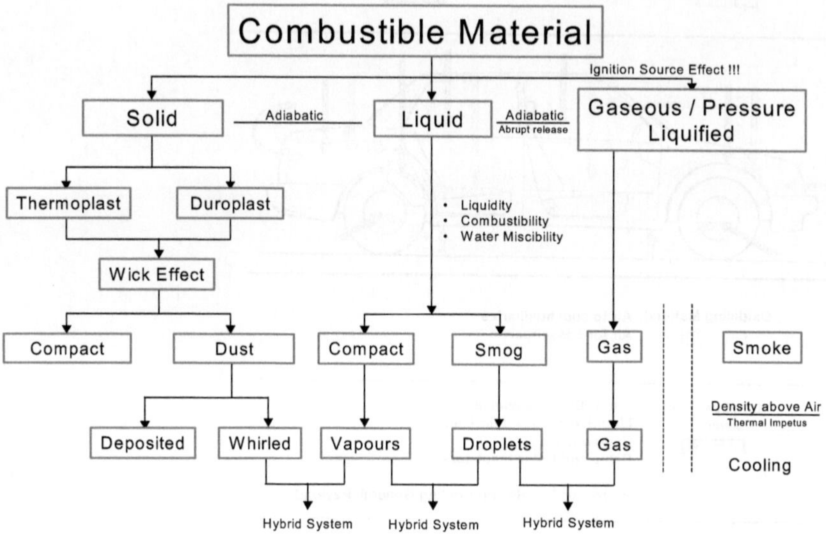

Fig. 4. Types of Fuel

2 Fire in an Automobile

When the EMMONS's Ignition Model is used for automobile fires, three areas of risk are present in a vehicle in totality, on the basis of which – with differing probabilities of occurrence – the basic criteria for the ignition maybe discussed (Fig. 5.):

- Luggage compartment (boot), if it is spatially separated from the passenger compartment
- Passenger Compartment
- Engine Compartment, where the external zones consisting of tyres are assigned to the engine compartment for a simplification of the model; one must note here, that fires resulting from tyres play a significant role only in commercial vehicles (approx. 16% of fire sources), while they play almost no role in passenger vehicles (approx. 0,6%, Fig 8.)

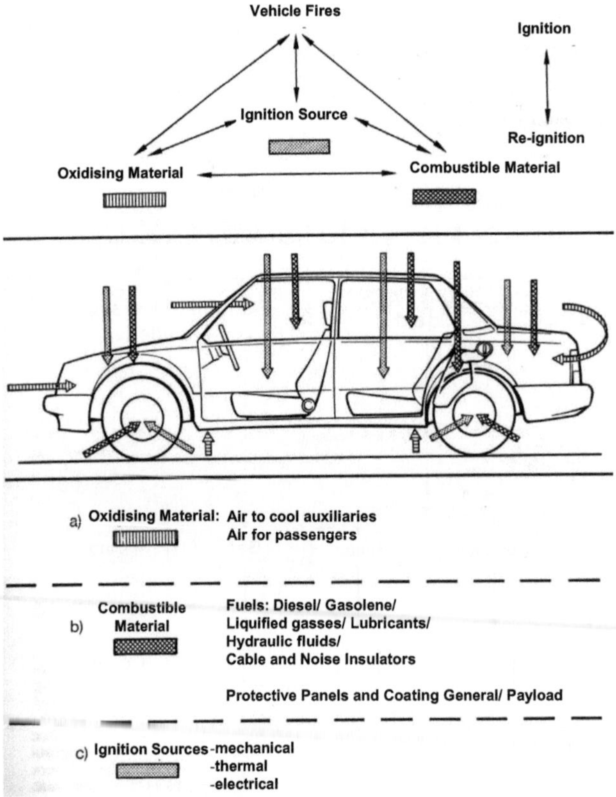

Fig. 5. Automobile fire from point of view of Emmons's ignition tetrahedron

Unfortunately in the area of automobile fires, only very little isolated statistical material is available. Hence one must fall back to the old BASt-Study from 1989 (NRW), which is shown in Fig. 6., Fig. 7., Fig. 8. This investigation showed (Fig. 8.), that the cable fire which is associated with approx. 30% of ignition sources, has a substantial significance if one additionally considers that a substantial proportion of ignition sources "Breakdowns in the Electrical Powernet" could be included in the areas of "Carburetor Fire" and "Engine Fire".

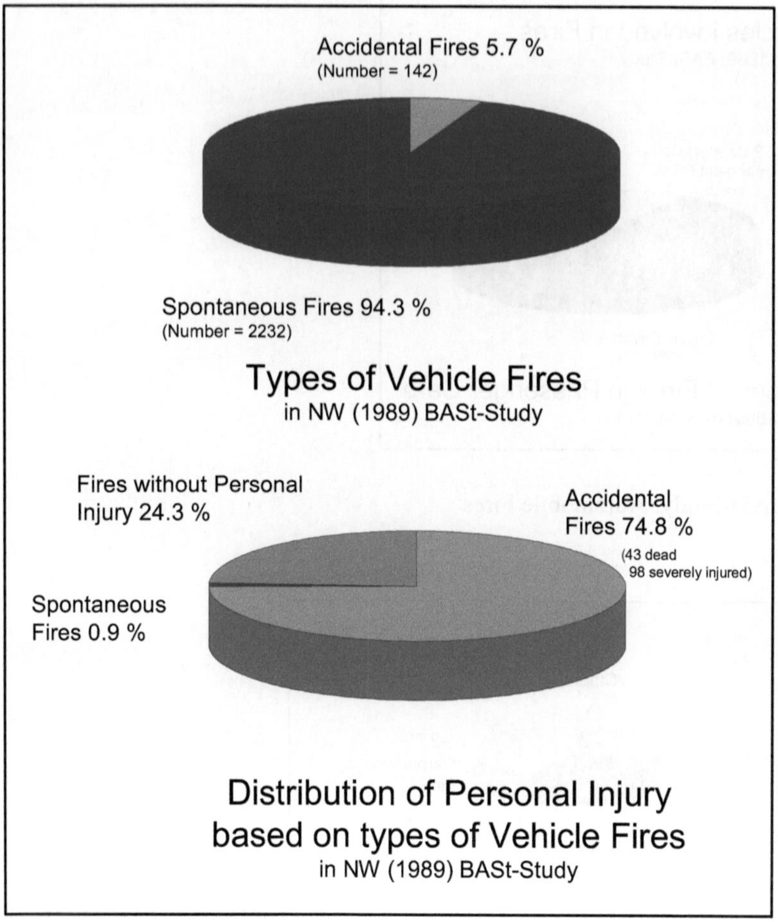

Fig. 6. BASt-Study: Automobile Fires

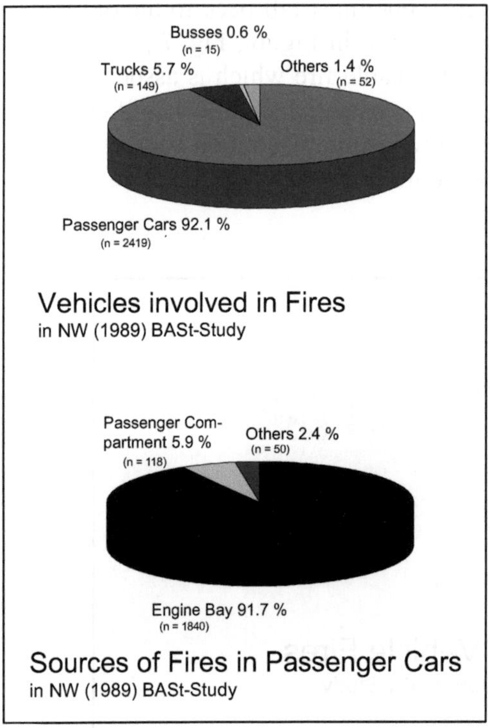

Fig. 7. BASt-Study: Automobile Fires

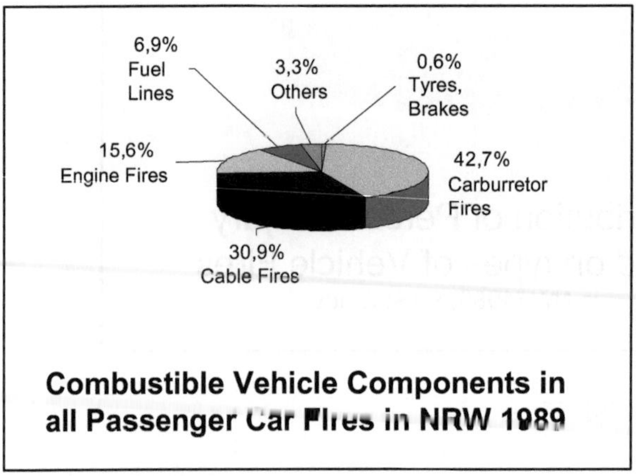

Fig. 8. BASt-Study

3 Cable Fire as a source of Fires in Automobiles

Based on the statements made so far, the overheating of the electrical network in the vehicle as a result of breakdowns is seen as the essential source of a subsequent fire. What is most disturbing in this context is that these breakdowns and the consequential damage can take place both when the vehicle is in operation as well in the parked condition. This is possible due to the fact that the electrically operated components have also to be operated using the powernet over the "continuous power supply function" in the parked position of the ignition system. Usually one speaks of areas of the electrical network which are protected against overload by suitably designed fuses. However, due to insulating defects in cables 50 etc., assembly defects in manufacture, in this case, the wiring of the electrical network, particularly when subsequent changes in the cabling is undertaken for the installation of additional electrical systems, a mismatch in the consideration of the safety principles or wrong assembly can, in the extreme case, lead to vehicle fires. These cases are presented at the end of this document. Firstly however, the results of a few laboratory investigations are introduced which provide information on the temperatures one could expect along with the resulting fire hazards, if areas of the electrical network are overloaded.

In this case, one directly implies the burning of the cable insulation as well as vehicle components present in the area of exposure - here in particular the floor mats.

Additionally, in a further project, the probability of a deposit of dust on an electrical component leading to a fire in a vehicle as a result of accumulation of heat is studied.

3.1 Fire Hazard as result of thermally stressed Cable Insulation

The insulation of electrical cables is made up of various organic Polymers. Based on the differing technological demands made, different combustible compounds, mostly thermoplastic in nature, are used (Tabel 1., Tabel 2., Tabel 3.). Electrical cables are classified in different groups based on their chemical consistency, dimensions as well as their thermal capacity according to DIN ISO 1629, EN ISO 1043 and DIN 76 722 and may be identifiable using codes corresponding to the specified standards. Examples of the technical properties of the insulation of electric cables used in automobiles is summarized in Table 4. and Fig. 9. presents a list of the maximum temperatures (long-term operating temperatures as well as peak temperatures) to which these insulating materials can be heated, supplemented by the properties two materials in the highest temperature class (260°C) as shown in Tabel 5.

The block of codes is made up of two groups:

1. Type code
2. Constructional code and material code

The sequence of codes used in the code block describes the construction of the cable from inside to outside.

Table 1. Keys for type codes used to represent electrical cables

Code	Description
FL	Low voltage cable according to DIN 76722 for road vehicles
FZL	High voltage ignition cable for road vehicles

Table 2. Constructional Code

Code	Description
W	Resistor cable or resistor core (in resistance ignition cables)
M	Mixed conductor or other conducting materials such as electrolytic copper or resistance alloys
F	Plain cable
Z	Multi-core, separable cable
R	Wall thickness of the insulating cover reduced
B	Film shield
C	Copper wire mesh
D	Copper wire protection
G	Glass fibre mesh
L	Painted
P	Insulating film
T	Textilumflechtung
Sn	Copper plated with tin
Vn	Copper plated with nickel

Table 3. Material codes

Code	Description
Y	PVC
YK	PVC Low temperature resistant
YW	PVC heat resistant
X	PVC meshed
2Y	PE
2X	PE meshed
4Y	PA
5Y	PTFE
6Y	FEP
7Y	ETFE
9Y	PP
11Y	PUR
51Y	PFA
G	NR
	SK
	SBR
2G	SIR
3G	EPR / EPDM
4G	EVA
5G	CR
6G	CSM
53G	CM (PE-C)
33X	BETA®-HX
41X	BETA®-ZX
31Y	TPE-SEBS
31Y	TPE-PEE

® Registered Trademark of Draka Deutschland

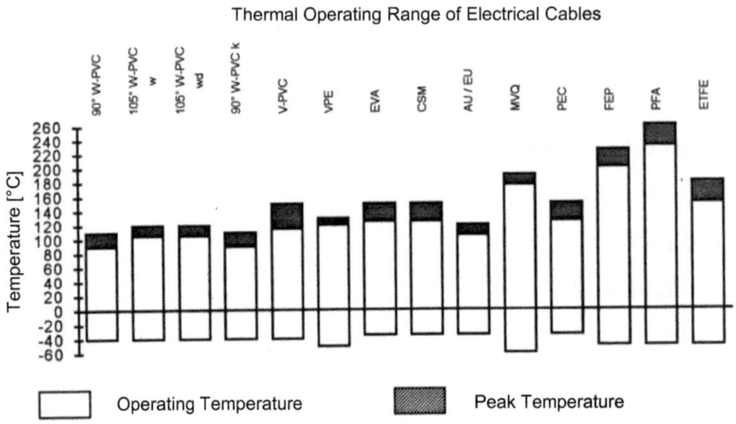

Fig. 9. Thermal operating range of electrical cables

Ms. D. Windhövel first investigated the temperatures and the resulting fires in insulating materials as a result of overload (currents >> 100 A) on various low voltage cables in one of her theses ("The Electrical Equipment in an Automobile as an Igniting Source", BUGH W., SA 9.93).

Fig. 10. Board to Control the Current

Table 4. Mechanical, thermal and chemical properties of insulating and coating materials

Nomenclature				Properties (Standard Values)									
				Mechanical			Thermal			Chemical Resistance (Standard Values)			
Code	Chemical	VDE	allowable Operating temp. acc. to VDE °C	Tensile Strength N/mm²	Strain %	Abrasion Behaviour	Low Temp. Behaviour	Flame Retardance	Corrosive Gases in case of Fire	Oils Greases	Sol-vents	Diluted Acids - Bases	Water 70°C
	Thermoplastics												
PVC	Polyvinylchloride Mixtures	Y	70-105	12,5-25	125-350	medium - good	moderate –good	medium – good	Hydrogen Chloride	mod. - medium	mod.	good	medium – good
PA	Polyamide	4Y	80	50-60	50-200	very good	good	good	-	very good	good	very good	medium
PUR	Polyurethane	11Y	80	35-50	500-700	very good	good	moderate - medium	-	good	good	moderate - medium	medium – good
	Elastomers												
NR SBR	Natural Rubber Styrol-Butadien Rubber Mixtures	G	60	5,0-10,0	300-600	moderate – medium	very good	bad	-	bad	bad	medium	medium – good
SIR	Silicon Rubber	2G	180	5,0-10,0	300-600	moderate	very good	moderate – good	-	good	bad	moderate	very good
EPR	Ethylene-Propylene Polymer Mixtures	3G	90	5,0-10,0	300-500	moderate – medium	good	moderate – bad	-	mod. - medium	mod.	good	very good – good
EVM	Ethylene Vinylacetate Copolymer Mixtures	4G	120	8,0-12,0	200-350	moderate – medium	good	moderate - medium	-	mod. - medium	mod.	medium	good - medium

Table 5. Temperature class 260°C

Code DIN ISO 1629, DIN 7728	PFA	PEEK
Code DIN 76722	51 Y	
Explanation	Perfluoraloxy-Copolymer	Polytherketone
Typical applications in automobiles	Brake cables	Brake cables
Used by	Various automobile manufacturers	
Advantages	General properties	General properties Low temperature flexibility
Disadvantages	Price	Price
Remarks	Contain halogens Non-combustable	Halogen free Non-combustable

Ms. Windhövel used a simple arrangement for these series of investigations, consisting of a 12 V-Starter battery, a board with two clamps for fastening the battery main switch and the section of the cable to be tested. The apparatus also includes measurement devices to test the voltage and the current through the cable, as well as a temperature measuring unit with a registered PC connection. A calculated maximum variable current of approx. 200 A, which could drive a maximum of 42 vehicle headlights as consumers, is provided (Fig. 10.). Each of the tests was recorded on video.

Firstly the temperature characteristics of the cable cores are measured for varying cable cross-sections with identical insulating materials (0.5/0.75/1.0/1.5 and 2.5 mm^2; FLRYA) and high, varying currents (Fig. 11.). The results show that the cables with low cross-sections (0.5 – 1.0 mm^2) burn relatively rapidly when subjected to high currents (approx. 12 – 20 sec.), the highest wire temperatures being around 500 - 800°C. The two cables of large cross-section (1.5/2.5 mm^2) however, burn only at very high currents (approx. 180 A) with a maximum temperature being 800 - 1000°C (Fig. 15.). A comparison of the temperatures of the wire, isolation and clamps (Fig. 16.) show that, as before, a comparable current flow through the wire led to the highest and most uniform temperatures, resulting in this measurement point being retained.

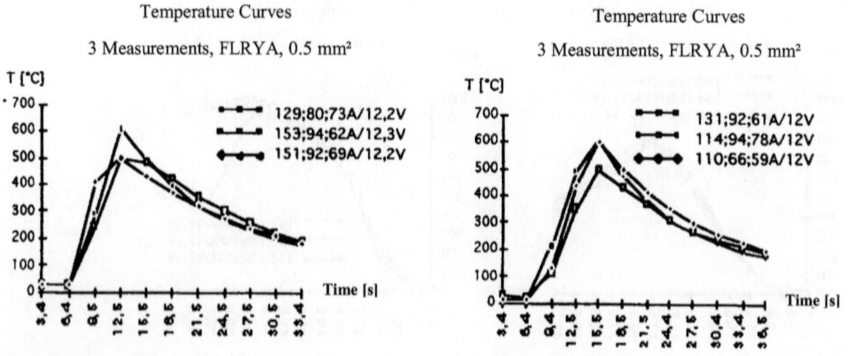

Fig. 11. Cable 0.5 mm²

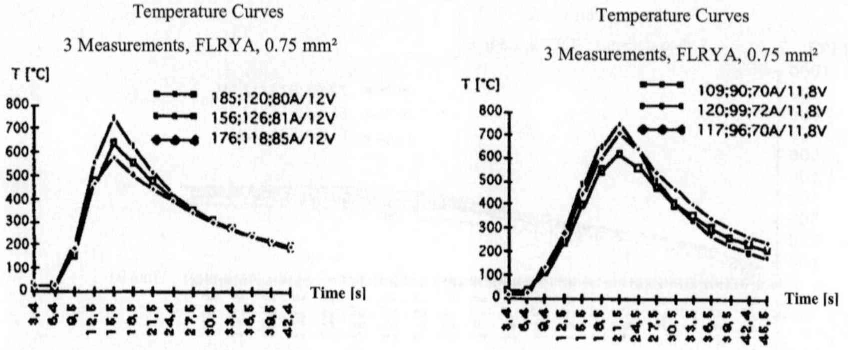

Fig. 12. Cable 0.75 mm²

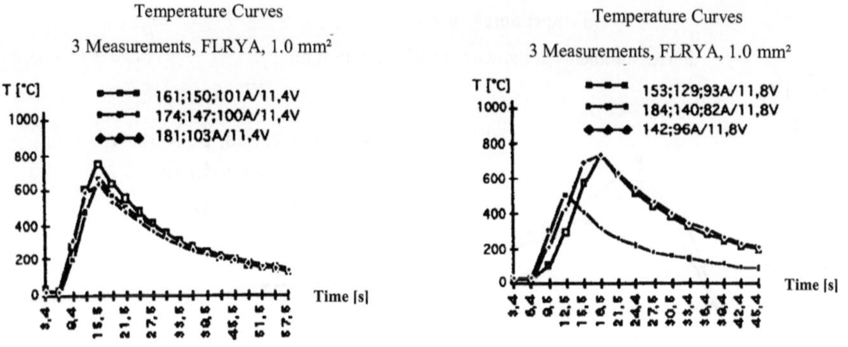

Fig. 13. Cable 1.0 mm²

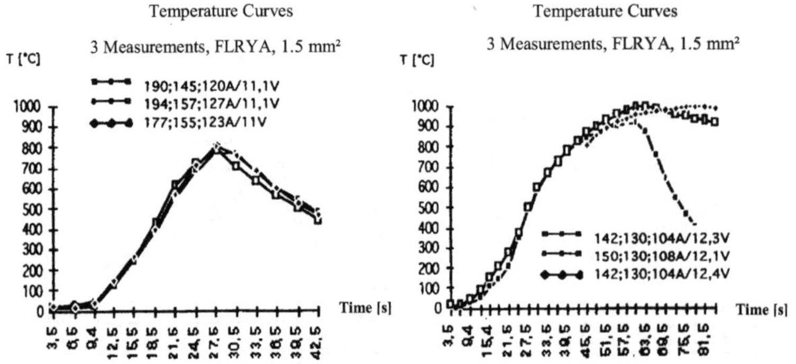

Fig. 14. Cable 1.5 mm^2

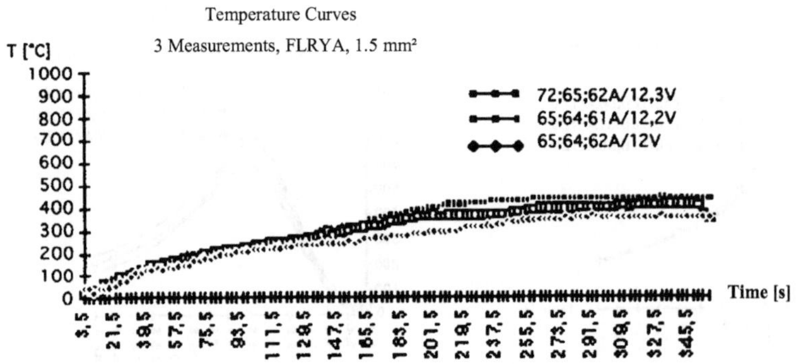

Fig. 15. Temperature curves for the 1.5 mm^2 cable

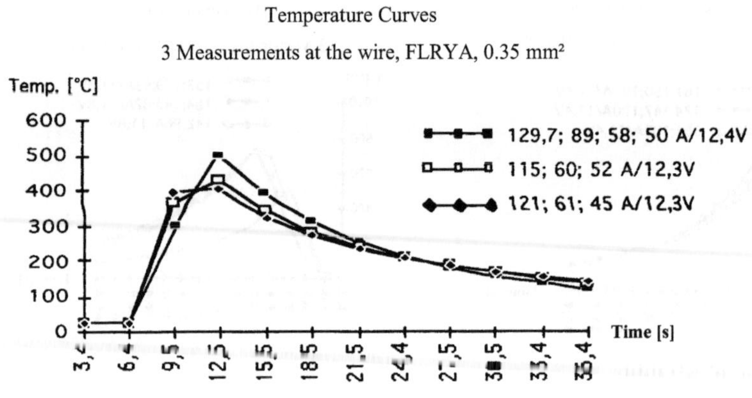

Fig. 16. Temperature curves at the wire

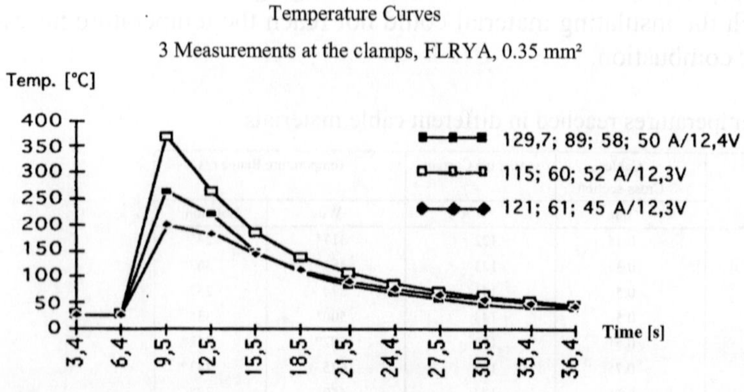

Fig. 17. Temperature curves at the clamps

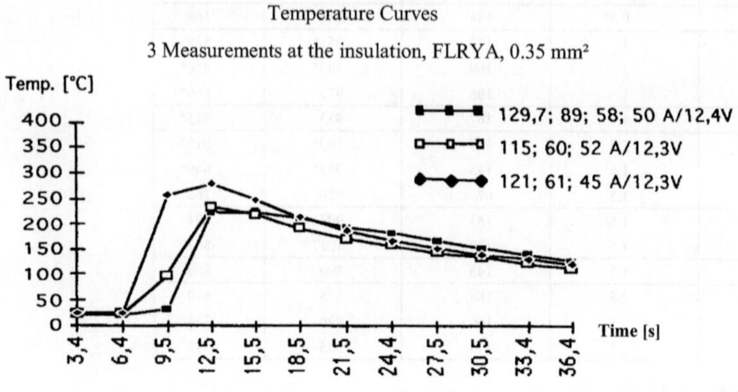

Fig. 18. Temperature curves at the insulation

Table 4. lists the significant experimental data, which can be summarised as follows:

- Cables with smaller cross-section (< 1 mm^2) burn rapidly at the given high current, such that the wire reaches clearly lower end temperatures (400 - 800°C) when compared to cables of larger cross-section (800 - 1000°C).
- Even though the geometry of the experimental cable used in this series of tests was unfavourable to combustion, the wires which were heated and allowed to hang freely in atmospheric air caught fire, at a current of 180 A when firstly the insulation FLRYA (2.5 mm^2) was used, and secondly in practically all experiments carried out using the insulation FL2G (SIR; 0.75/1.5 mm^2; Fig. 19.). In the other cases, the insulation

melted rapidly and fell out of the freely hanging cable core as a result of which the insulating material could not reach the temperature necessary for combustion.

Table 6. Temperatures reached in different cable materials

Insulation	Cable-Cross-section	Average Current	Temperature Range / °C	
	/ mm²	/ A	Wire	Isolation
FLRYA	0.35	122	417*	245
	0.35	123	483*	363*
	0.5	141	574	257
	0.5	144	500*	335
	0.5	118	602*	252*
	0.75	167	625	443*
	0.75	172	650	373
	0.75	115	687	423*
FL2G	0.75	155	908	552*
	0.75	124	886	622
FL6Y	0.75	132	929	851
Arnitel	0.75	144	585	569
FLRYA	1	172	653*	525
	1	160	700*	450*
	1	106	973	869*
Arnitel	1	167	955	823*
FLRYA	1.5	187	780*	519*
	1.5	145	993*	649*
	1.5	67	426	262
FL2G	1.5	183	907	694
	1.5	127	1027*	867*
Arnitel	1.5	143	890	800
FLRYA	2.5	183	958	810
	2.5	160	890	754*
	2.5	114	583	545*

* only 2 out of 3 experiments averaged

Fig. 19. Burning insulation FL2G

3.2. Fire Hazard due to the Internal Coating on an overloaded Automotive Electrical Cable

In this thesis (Th. Oertel: „The Electrical Cables in an Automobile as Igniting Sources"; BUGH Wuppertal; SA 07.97) one investigates whether an overload on electrical cables could lead to neighbouring or adjacently located combustible components such as the passenger compartment to be set on fire. Hence it is to be determined if low voltage cables of varying dimensions with varying insulating materials can set their direct surroundings - in this case a commonly used automobile floor carpet - on fire when subjected to a definite electrical overload.

The experimental apparatus consists of the same table as shown in consisting of 2 connecting clamps and the battery main switch, a 12 V Starter battery, the board consisting of 40, 55 W headlamps as consumers, a Quartz tube, 100 cm long and 3.6 cm in diameter with 2 fixing points (from a Ring oven conforming to DIN 53 436 for thermally stressing textiles) and 2 connecting clamps, each with 100 cm connecting cables (section 10 mm^2). Current and voltage are, as before, displayed and recorded digitally. The quartz tube serves to determine failure due to thermal or electrical overload on the respective wire sections of length 40 or 80 cm as well as to create a reproducible environment for the test.

Additionally, the cylinder – when needed – can be cleaned of condensate or other coating simply by annealing in a Ring oven.

Commonly used automotive electrical cables of varying section with three types of insulation are used as experimental samples:

- PVC-Insulation FLRY in sections of 0.5, 1.5 and 2.5 mm^2 with a reduced insulation wall thickness;
- Cables from the TPE group with the name ARNITEL (TPE-E) in sections of 0.75, 1.0 and 1.5 mm^2 where the wall thickness of the insulation is again similarly reduced;
- BETAX (Poly-Olefin-Polymer / Polyester) Cross-section 0.75 mm^2

Samples of the above mentioned cables are inserted into the quartz tube (of respective lengths 80 or 40 cm) which is enclosed using cork stoppers and subjected to thermal stresses by differing overloading currents and the melt down time is first determined (Fig. 20.).

The following phenomena could be monitored during the process of overloading the cables:

- inflation of insulation,
- softening of insulation,
- melting of insulation,

- formation of blisters,
- development of smoke,
- separation of insulation from the wire,
- glowing of the wire,
- melt down of the wire and
- detonation of the smoke.

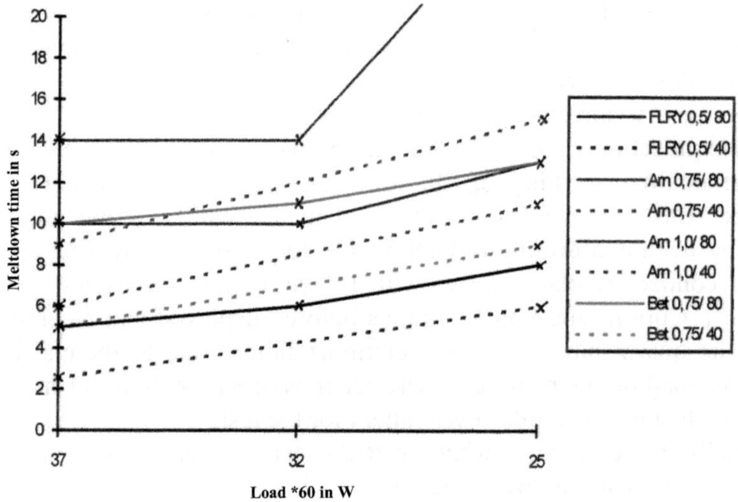

Fig. 20. Dependence of melt down times on load

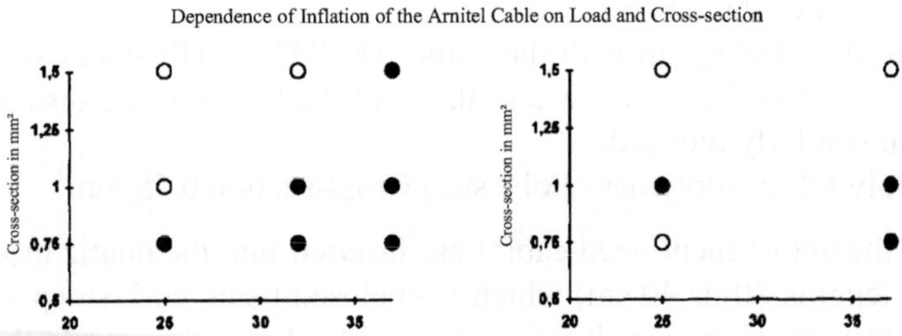

Fig. 21. Igniting different Cables a) 80 cm, b) 40 cm long

The earlier mentioned phenomenon of an ignitable oxidative conversion of the pyrolysis gases could be monitored in the ARNITEL-Insulation of both cable lengths (Fig. 21.). The PVC-insulated cable did not show such a phenomenon in any of the experiments, while in the case of the BETAX-

Insulation, an ignition of the pyrolysis gases or vapours took place for both cable lengths only at the highest loads.

Fig. 22. now shows the main experiment carried out with a 10 x 20 cm section of the floor mat and a 25 cm long cable sample as well as the experimental process using a highly loaded 1.5 mm^2 ARNITEL-Line.

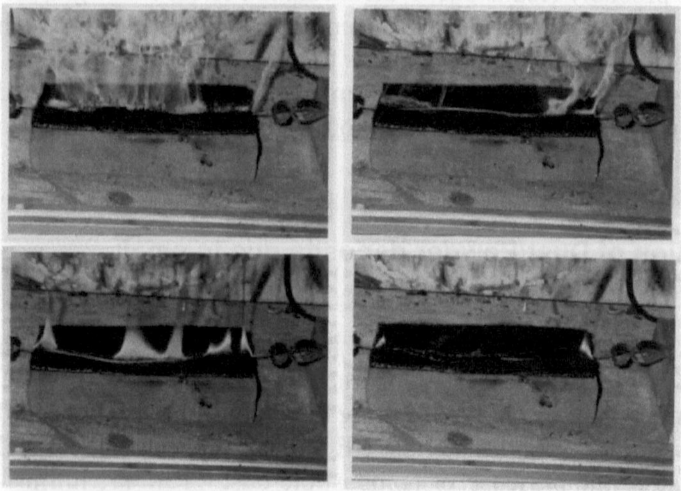

Fig. 22. ARNITEL-Cable under Stress

Fig. 23. provides an overview of the results of these experiments. It shows that the ARNITEL-Insulation in its weakest form ignites all floor mat samples, though the ignition is displaced in the direction of high overload with increasing cross-section of the cable.

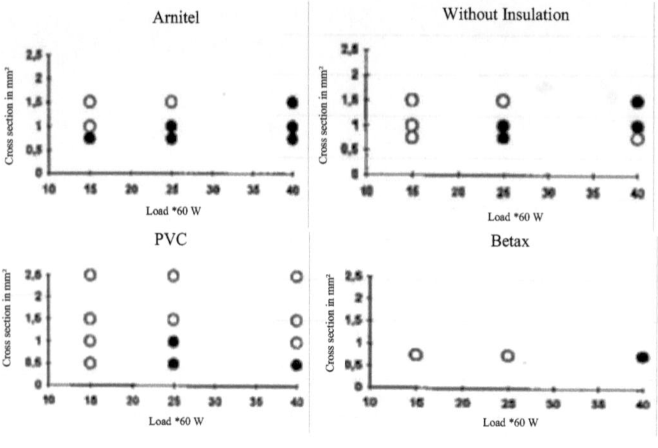

Fig. 23. Cables under Stress

The wires without insulation were expected to behave in a manner similar to the cable with the PVC-Insulation according to the results of the preliminary test, since the pyrolysing gases of PVC cannot bring in the detonating energy from the fluorescing cable core in addition to thermal energy into the ignition process. The absence of ignition in the larger PVC-insulated cables probably results due to the insulation effect of the graphitised supporting framework and the consequentially large distance between the fluorescing wire from the carpet which is to be ignited.

The BETAX-insulated cable shows the same behaviour as demonstrated by the preliminary experiment. The ignition of the floor carpet is also achieved here only at high currents.

3.3 Inflammation of Dust/Powder deposits due to small electrical components

The series of experiments introduced briefly in this section was undertaken by F. Borgert (DA 05.01; BUGH Wuppertal) in cooperation with DMT Dortmund and refers to the context of testing the draft of the IEC-Standard for intrinsically safe operating materials (IEC 61241-11, Ed.1; Electrical apparatus for use in the presence of combustible dust - Part 11; Intrinsically safe apparatus "iD"). Electrical resistances of differing powers and geometry (Table 7.) are introduced into dust fills (Table 8.) in a defined manner. Combustible samples of dust with combustion numbers between 1 – 5 are used (Table 7.).

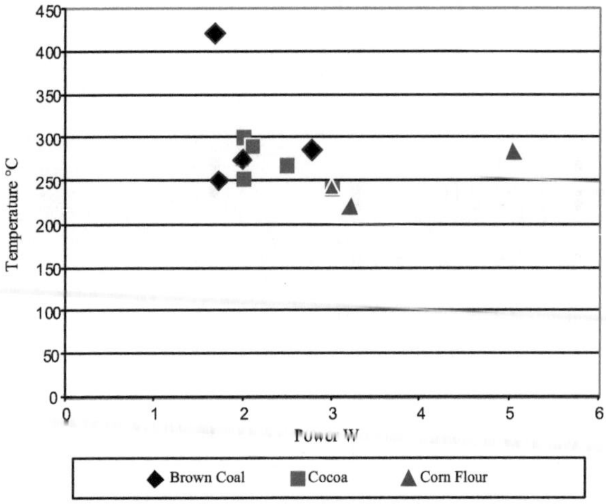

Fig. 24. Power in experiments with Inflammation

Table 7. Electrical resistance of differing powers and geometry

No.	Resistance Specifications	Electrical Resistance [Ω]	Nominal Power [W]	Form	Measurements [mm]	Surface Area [mm^2]
a	$R_{0,125}$1200	1200	0.125	Cylinder	l * d 3,5 * 1	13
b	$R_{0,25}$100	100	0.25	Cylinder	l * d 5 * 2	38
c	$R_{0,6}$220	220	0.6	Cylinder	l * d 5*2	38
d	$R_1$120	120	1	Cylinder	l * d 11 * 3,5	140
e	$R_2$470	470	2	Cylinder	l * d 11 * 5	212
f	$R_5$68	68	5	Cuboid	l * b * h 25 * 6,4 * 6,4	722
g	SMD220	220	Unknown	Cuboid	L * b * h 3 * 1,5 * 0,5	13

Table 8. Safety codes

Dust	Density [g/cm³]	Median [µm]	Moisture Content[1] [Weight-%]	Min. Ignition Temperatur [°C]	Combustion Behaviour
Brown Coal[2]	0,55	38	10,3	230	BZ 4
Cocoa	0,375	27	3,66	240	BZ 4
Corn Flour	0,45	155	14,37	> 450[3]	BZ 5
Aluminium Dust[4]	0,3	36	0	230	BZ 4
Detergent (ja-Delicate Washing Powder)	0,8			Detergent without detailed specification 380 - 390[5]	BZ 3
Detergent (Persil Megaperls)	0,75				BZ 3
Detergent (dalli Stain Remover)	0,7				BZ 1
Sulphur – Brown Coal Mixture 50 / 50		31	12,34	Sulphur 250 - 280	BZ 5
Sulphur – Brown Coal Mixture 25 / 75		35	11,32		BZ 4

[1] Moisture content defined at 105°C
[2] Codes are already defined by DMT under the sample number 98057
[3] No fluorescence at 450°C
[4] Codes are already defined by DMT under the sample number 92034
[5] Values lt. BIA_Report, Combustion and Explosion parameters of powders, Nov. 1977

Table 9. Data overview

Dust	Ignition Temperature [°C]	Min. Ignition Temperature [°C]	Combustibility [Combustion Number]
Aluminium in powdered form	500 - > 850	360 - > 450	1 – 4
Brown coal deposited dust	450 – 480	250 – 300	4
Buchenholz	400 – 500	310 – 320	4 – 5
Cocoa varying fat content	490 – 590	250 – 260	4
Sulphur	240 – 370	250 – 280	5
Corn dust[1] (maize)	580	460	3

[1] Values lt. BIA_Report, Combustion and Explosion parameters of powders, Nov. 1977

The systematic experiments showed the following surprising results:

- the minimum power at which one could ignite a filling of dust was 1,7 W,
- the general power required for the ignition of the investigated system was in a range between 2 – 3 W (Fig. 24.),
- the surface temperature of the resistances used for the ignition was in the range between 250 - 300°C and
- a power density of approx. 10mW/mm² was in most cases already sufficient for ignition.

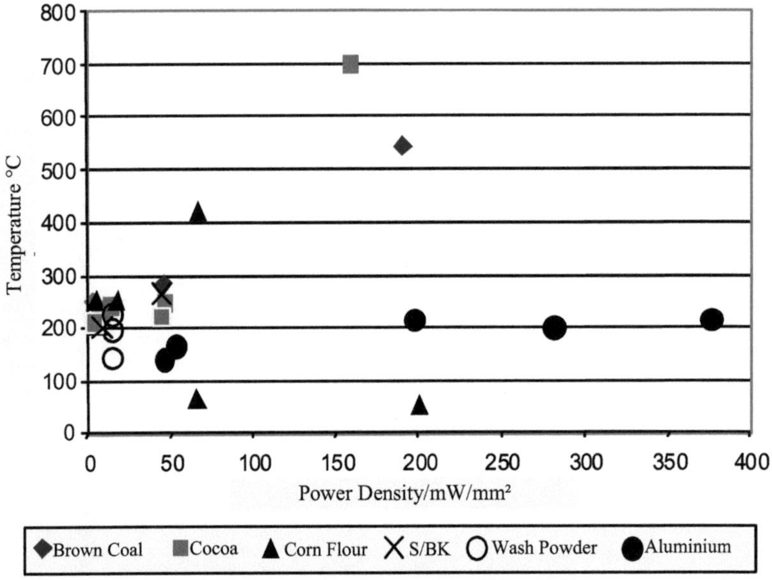

Fig. 25. Power density in experiments without Inflammation

4 Summary

The electrical powernet in a vehicle always represents a latent and potential fire risk. Hence a number of theses and dissertations have been undertaken in the Faculty of Safety (Faculty 14), in the area of specialisation „Fire and Explosion Protection", the results of which have been presented in a summarised form in this report.

In the study thesis introduced at first (D. Windhövel, SA.; 05.93), the temperature development in vehicle electric cables of varying dimensions and insulation was systematically tested under the influence of overloading currents. The results showed that cables of small cross-sections (< 1 mm^2) rapidly caught fire and burned under the influence of the specified current (upto 200 A), the overloaded wire however reaching lower maximum temperatures (approx. 400 - 800°C) as compared to those with larger cross-sections (800 - 1000°C).

The cable with the code FL2G (Silicon rubber SIR), was set on fire regularly and with certainty in all experiments, while from the remaining materials, only the insulation FLRYA of section 2.5 mm^2 was set on fire at a current of 180 A in. Since the experimental cables were allowed to hang freely, a phenomenon that occurred with high probability in these series of experiments was the fact that the insulation melt of low viscosity dropped away from the wire and hence could not be thermally stressed for longer periods of time.

In a further thesis on the same topic (Oertel, Th.: The Electrical Cables in Automobiles as Igniting Sources; SA; BUGH Wuppertal; 07.97) different experimental parameters were specified, more oriented towards the pyrolysing conditions in the installed position of the cables in the vehicle. Firstly the melt down time of cables of 40 cm and 80 cm length respectively was determined using an enclosed quartz tube. The results showed that doubling the testing length also doubled the melt down time (average 5 sec./40 cm; 10 sec./80 cm). The cables insulated with ARNITEL (TPE , Thermoplastic Polyether-ester-copolymer) showed rapid combustion reactions in the presence of the pyrolysis vapours, both in lengths of 40 cm as well as 80 cm.

The products of pyrolysis in the PVC-Insulation (FLK/FLRY) did not catch fire.

The BETAX insulation is a Polyester- or a Polyolefin-Copolymer. This insulation also produces a rapid reaction of the vaporised pyrolysis gases at high currents. The ignition tests on the floor mat sample shows that with increasing current, the possibility of ignition clearly increases.

The last section represented a summary of the results from a diploma thesis by F. Borgert (DA 05.01; BUGH Wuppertal). It deals with the possibility of ignition of dust deposits by small electrical components (resistances). Powers between 1,7 – 3 W already lead to surface temperatures between 250 – 300°C due to the heat accumulation as a result of a coat of dust over the component. The consequence of this heat accumulation was the ignition of the dust deposit, this being a well known problem in the operation of commercial vehicles (especially in agricultural use).

Circuit Protection Components for Future Vehicle Electrical Systems

Jo Jaspar

Engineering Manager Automotive, Littelfuse BV, Utrecht, Netherlands

Neil McLoughlin

Technical Marketing Manager, Littelfuse Ltd, Dundalk, Ireland

Patrick Bellew

Senior Test Engineer, Littelfuse Ltd, Dundalk, Ireland

Abstract

Future vehicle electrical systems functioning on a voltage of 42 – 48 V, have been discussed since the end of the 1980's; this is in anticipation of the need for greater power which is too limited with the 12 V system. During the second half of the 1990's, the "green environment policy" asking for a reduction of emissions received increased attention, plus vehicle safety related issues. This paper addresses challenges for 42 V systems and potential challenges of incorrectly specified circuit protection components.

1 Introduction: Higher system voltage, why and when?

Per Ref's [1] and [2], system voltages higher than 12 V were already discussed prior to 1991, anticipating present day automotive electrical systems reaching the limit of their efficiency. With the recent strengthened focus on:

- a reduction of CO_2 emissions
- an increase in vehicle performance and safety via information and navigation systems [3]
- a further decrease of vehicle manufacturing and warranty costs

the limits of present day systems are constantly being re-evaluated.

One may question the need for a 42 V system if only navigation/infotainment [4] is to be connected; up-scaling to 24 V proven technology would give savings on cable cross sections immediately, and allow a peak usable power of about like 300 A – 500 A x 24 V = 7.2 – 12 kW, depending on the cable size. This is about 3 – 6 times the today's peak power for large cars (2 kW). Increased short circuit protection could be reached using e.g. LF 500 A MEGA protector, rated 32 V/2000 A/2.5 msec.

As some predictions include future electrical power demand as high as 30 kW peak, forums such as MIT [5] in USA and SICAN in Europe [6] have been focussing on the 42V-PowerNet. Note that this is all new technology [7], [8], [9], [10], [11], [12] for which no history or track record exists. A quantification of such topics as Arcing, Aging [31] and Corrosion phenomena are yet to be explored. Nevertheless the first 42 V production vehicle Toyota Crown has been put August 20, 2001 in series production on the road in Japan [5] and [12], and includes the Mild Hybrid technology similar as to the Toyota Prius, reported in [4].

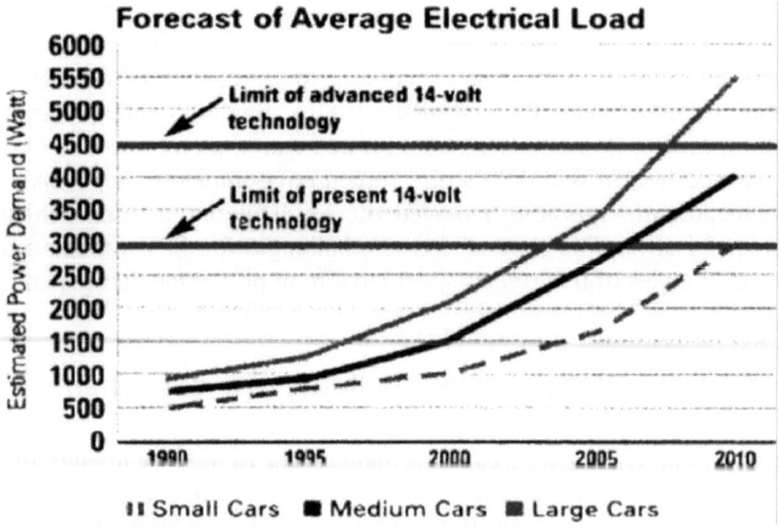

Fig. 1. Estimated in-vehicle electrical power demand

2 System Definition

The 42 V System definition includes technologies selection and parameter definition for Battery technology (lead-acid, nickel-metal-hydride, lithium-based), Integrated Starter-Alternator (various concept exist [13]), Regenerative braking (present or not; if so, storage in battery or ultra-cap), Semiconductor (break down voltage), Data-bus system (various protocols exist) and Drive by Wire. Furthermore it requires Level of Safety requirement:

- setting by the AOEM and
- determination and verification by the suppliers.

2.1.1 Over-current [System definition]

Assuming Lead Acid technology [14] (e.g. Toyota introduction and German Draft specification for 42 V Bordnetz), then increasing the battery voltage with a factor 3, will reduce currents a factor 3. It allows battery internal resistance, cabling and end-user resistance to go up a factor 9. If so, to-days 6 cells battery range 12 V/20 – 110 Ah/5 – 20 mOhm internal resistance is expected to transform, in first approximation, into 18 cells 36 V/6 – 35 Ah/45-180 mOhm internal resistance.

In addition:

- the integrated starter-alternator combination is able to deliver an excessive current. This current depends on the number of revolutions during driving, and of the capacity of the unit in kilowatts. The integrated starter-alternator unit is likely to be equipped with internal electronics limiting the output voltage to 58 V DC.
- there may be a regenerative braking unit on board. The energy may be stored either in a battery or an Ultracap. The higher the voltage, the smaller the size of the device will be. If this unit is connected to the 42V-PowerNet, it is able to deliver a further discharge or leakage current. This current depends on the mass and speed of the vehicle, the size in kilowatts of the unit, and the size of the storage battery/capacitor in question.

Some estimates for system current and voltage output have been illustrated below. Note that these estimates do not include:

- any onboard energy storage larger than experienced today.
- charging voltages for nickel-metal-hydride or lithium-based battery technologies currently in development
- details about maximum voltage and current output of the battery charger.

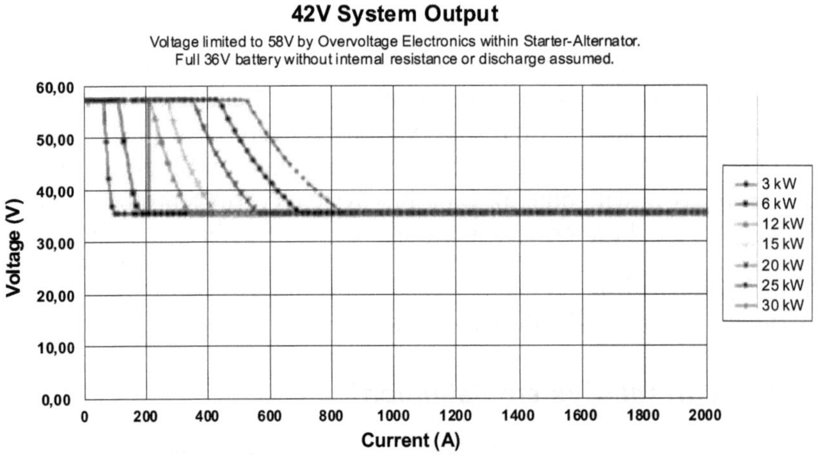

Fig. 2. 42 V-PowerNet Battery-Starter-Alternator Output example

When sealed lead-acid batteries come in highest stage of charging, internal re-combination causes temperature to rise slowly, which causes internal resistance to decrease. If appropriate limits are not used, this process becomes self-accelerating until damage to the battery cells occurs (excessive gassing, venting, heat distortion of the casing). Hence high battery charge currents for extended periods should be avoided because of thermal runaway risk.

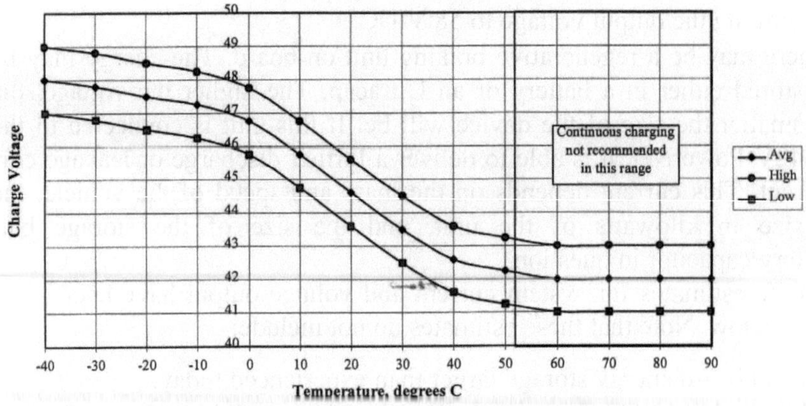

Fig. 3. Charge settings for lead-acid battery technology, as a function of ambient temperature [14]

Table 1. Short circuit current from battery

Maximum Short circuit current (A)		
	Battery Voltage (V)	
Internal Res. (mOhm)	36	48
45	800	1067
60	600	800
90	400	533
120	300	400
150	240	320
180	200	267

Table 2. Short circuit current from Integrated Starter Alternator

	Short circuit current (A)		
	Battery Voltage (V)		
ISA Power (kW)	36	48	60
3	83	63	50
6	167	125	100
12	333	250	200

Also, all electric end-users will have to be redesigned to accept the tripled voltage. At this time, little to no actual values are known for any of the end-users nor for the battery, integrated starter-alternator combination nor any of it's alternative solutions.

2.1.2 Over-voltage [System definition]

SAE J1113/11 issued June 1995 "Immunity to conducted transients on power leads" and ISO/DIS 7637 Parts 1 and 2 currently under evaluation, define for the 12 V/24 V system, a number of over-voltage pulses. These standards include:

- Pulse 1: disconnection of inductive load,
- Pulse 2: sudden interruption of current in parallel devices,
- Pulse 3: transients as a result of switching processes,
- Pulse 4: energising starter motor circuit and
- Pulse 5: disconnection of battery while charging (Load dump).

Older SAE J1113 issued 1987 and ISO 7637 Part 2 issued 1990 specifications describe in addition to the pulses listed above:

- Pulse 6: current interruption of ignition coil and
- Pulse 7: alternator field decay at turn-off.

which refer to alternators not having transient voltage suppression built in.

For classic 12 V/24 V systems, above mentioned ISO/DIS 7637 Parts 1 and 2 reports peak voltages up to:

- 150 V for 12 V system (equals 12.5 times 12 V) and
- 600 V for 24 V system (equals 25 times 24 V).

For the 42V-PowerNet, it is anticipated that many of the pulses defined for today's 12 V systems may be retained:

- showering arcs (pulses 3a and 3b) caused by relay chatter have been shown to be independent of the DC bus voltage [15].
- To maintain the same motor torque for example, the 3-fold reduction in current will necessitate a 3-fold increase in the number of windings. The inductance increases with the square of the number of windings. This will exactly compensate for the reduction due to the current–squared term, Thus:

$$W\ (\text{energy}) = 0.5 * L_{42V} * I_{42V}^2 = 0.5 * 9 * L_{14V} * (I_{14V}/3)^2 \tag{1}$$

Therefore:

$$W_{42V} = 0.5 * L_{14V} * I_{14V}^2 = W_{14V} \tag{2}$$

i.e. the transient energy will remain the same.

- the time constant, $L_{42V}/R_{42V} = 9L_{14V}/9R_{14V}$ will also remain the same.

However, it is to be noted that:

- Given that the higher voltage system and loads will have different parasitic inductance and capacitance, the actual pulse shapes may differ from the ones quoted in today's standards.
- In some publications there is a view that "showering arcs" (pulses 3a and 3b) due to relay switching, will be eliminated and replaced by solid state switching. This would result into redefining the transient environment. However there is also a view [35] that new relay designs, with dual series contacts and/or hybrid relays with in-built transient suppression, will replace the existing types and showering arcs, while changed, will still be present.

The proposed 42V-PowerNet load-dump waveform (Fig. 4.) is a suppressed waveform, i.e. it assumes an integrated Transient Voltage Suppressor (TVS) device. The 58 V maximum dynamic over-voltage specifically defines the transient output of the 42 V alternator, and accommodates the worst-case energy transient that drives the current requirements in 14 V systems for over-voltage protection.

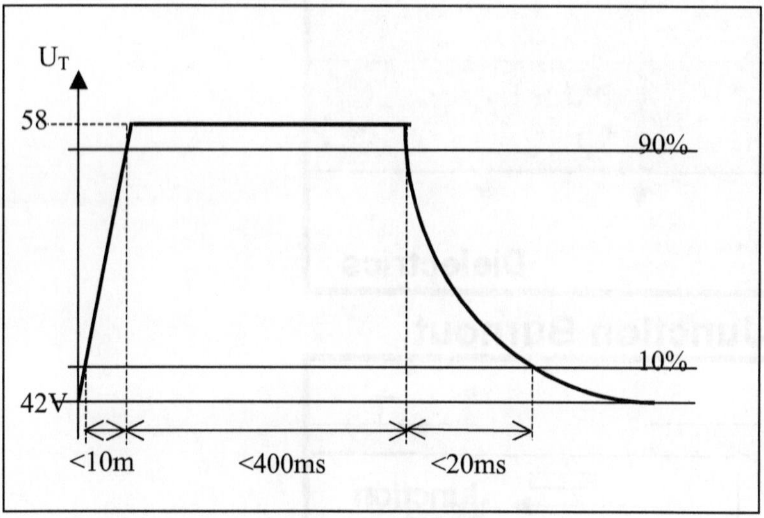

Fig. 4. Proposed 42V-PowerNet Load dump test pulse

Any other suppression devices fitted to the 42V-PowerNet must be over-specified in energy capability terms or have clamping voltages ***in excess of*** 58 V if they are not to share in the central suppression task. This as sharing would mean over-specification in energy capability terms. Further, a higher harness resistance, if realised, will aid to isolate modules, reducing the need for clamping voltage rise above 58 V. Accurate modelling of the load dump behaviour of ISA (including any of the starter-alternator output due to failures such as sudden battery disconnect, voltage regulator failure or short between rotor and stator in case of crash) may need to be carried out [16], [17], [18], [19].

As with the 14 V system, it is not expected that the central suppressor (now integral to the alternator) will provide the required suppression of these other pulses referenced in the standard. Given this, and the need to ensure that module protection does not function as part of the central suppression, it is anticipated that the protection level on "remote" modules will be greater than 58 V. Imposing a 58 V max requirement on the complete system will produce suppression requirements that may prove difficult to realize in a cost effective manner.

2.1.3 Electro Static Discharge [System definition]

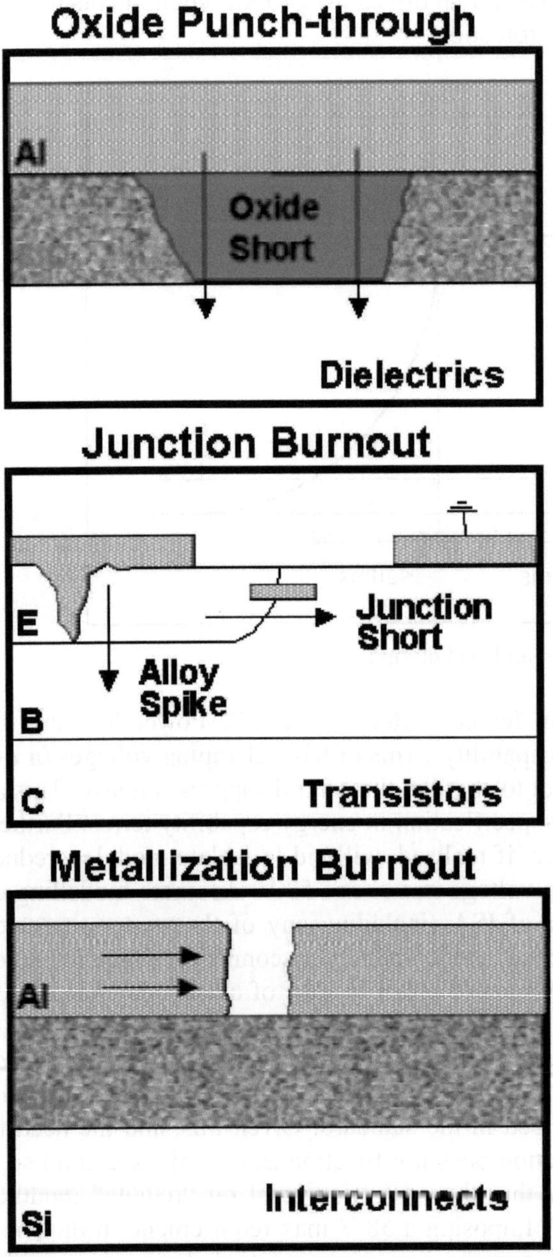

Fig. 5. Electric Discharge Failure Mode examples of Electronic components.

With reference to Immunity to Electrostatic Discharge (ESD) as per SAE J1113/13 Issued Feb 1995 revised Oct 1997 and ISO/DIS 10605 under discussion since 1999, we anticipate at first instance these pulses would not change due to a higher system voltage. As the "smart" electronic solutions rely more and more on size-reduced components and include ESD protection up to 2 kV, additional ESD suppression will still be required to ensure protection against high level (15 – 25 kV) ESD events.

Table 3. Table showing Silicon technology ESD vulnerability

Type	Value
EPROM	> 100 V
CMOS	> 250 V
BIPOLAR	> 350 V
TTL	> 300 V
SCHOTTKY TTL	> 1000 V

2.2 Challenges

2.2.1 Over-current [Challenges]

When a fuse interrupts, the current heats the fuse element until melting and subsequent boiling occurs, creating a metal gas. The system inductance prevents a current change to zero, and the energy stored in the system (inductance) is released in the arc. The metal arc gas is extremely high; temperatures during arcing can be such that air dissociates. It expands in a shock wave like characteristic. In this expansion it burns and carbonizes the surface layer of the plastic fuse housing.

As the plastic housing is relatively cold, the metal gas condenses onto the burned surface of the plastic. If the dielectric strength of this structure is sufficiently high, the voltage of the system can be held, and only a small leakage current flows via this layer. If the dielectric strength of this structure is not sufficient, it breaks down in time and the voltage of the system causes a large current to flow using the fuse housing as a conductive path. Due to this current and the relatively high resistance, the plastic housing may be ignited causing even more over-current to flow, and may ultimately lead to total system destruction.

The challenge arises in building components, capable of interrupting the short circuit current, maintaining a high insulation resistance over time once opened, and being as small as possible. Components may either be single or multi-element type of devices.

2.2.2 Arc Fault protection [Challenges]

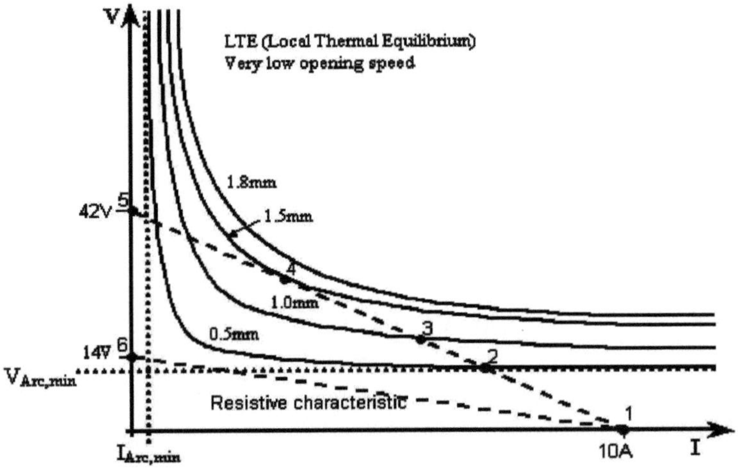

Fig. 6. Langmuir Arc V-I Characteristics example.

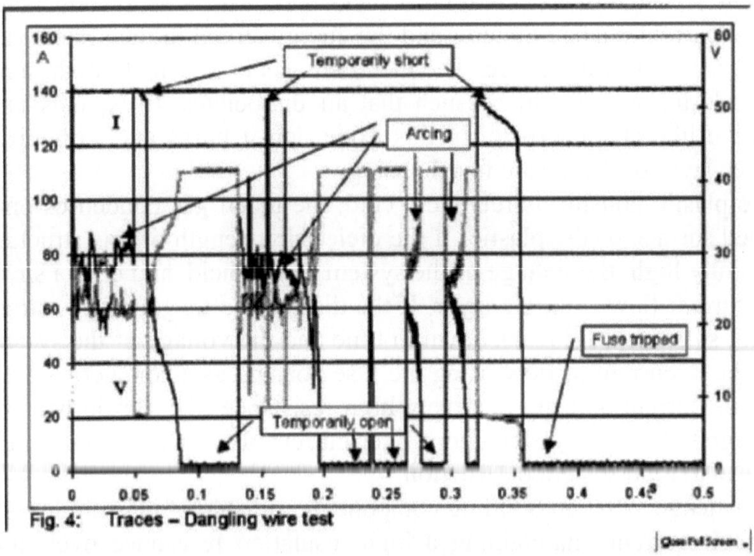

Fig. 7. Irregular current as a function of time during dangling wire test [8]

Per Langmuir for arcs in approximation $I^nV=C$, and hence by rising the system voltage a factor 3 (from 14 V to 42 V), the current at which stable arcing can occur (assuming same materials involved) is lowered. Do we see for 12 V systems, arcing only within the fuse, for 42 V – 58 V systems one may observe arcing anywhere in the vehicle, without interruption by the fuse element in a timely manner.

The challenge arises in building components, capable of detecting the presence of the arc-fault using the current profile as a function of time, and interrupting the circuit current within milliseconds. Component shall be a small as possible. At this time the market seems unclear, whether the component needs to be a resettable device, either manually or automatically.

2.2.3 Welding during insertion or extraction [Challenges]

Insertion or extraction of any clip type connection in an energised 42 V rated system, will give rise to welding effects and arcing, regardless of material selection of fuse and clip. Hence:

- When, during vehicle crash, splices inside harnesses are torn apart creating arcs, precautions need to be taken under bonnet gasses or fluids are not ignited.
- Fuse-blocks can no longer be accessible without voltage turn off prior to opening.

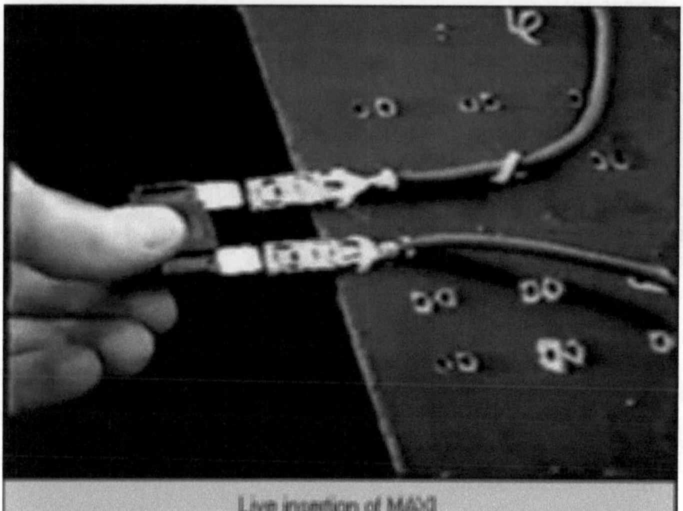

Fig. 8. Example of live insertion and extraction under electric load

The challenge arises in building components, capable of detecting the presence of the irregular current profile as a function of time, and interrupting within milliseconds the circuit current. Component shall be as small as possible. At this time, the market is unclear whether the component needs to be a resettable device, either manually or automatically.

2.2.4 The Dual voltage system for interim stage [Challenges]

Possibilities for dual voltages on board of vehicles to be considered are:

- 42 V/14 V dual voltage internal combustion vehicle, as an interim stage until 42 V technology is fully mature. Currently all AOEM's are believed to have prototype dual voltage vehicles for 42 V system testing available,
- hybrid internal combustion/electric engine/12 V, e.g. as in Toyota Prius and
- hybrid internal combustion/electric engine/42 V /12 V, e.g. as in Toyota Crown launched 2001 in Japan.

The high voltage system can completely destroy the lower voltage system by sustained overvoltage. Also, the 42 V battery may deliver such a high current to the 12 V system, that due to is 9-fold higher internal resistance, it may quickly overheat. In addition, SICAN reported possibilities of ejection of lower voltage battery fluid or even battery explosion.

The challenge arises in:

- keeping the voltages completely separated by packaging instructions,
- limiting the proliferation of highest voltage to a minimum,
- limiting the necessity of a common ground to a minimum as a common ground may allow an unintended closure of possible short current path,
- limiting the current flow from the high voltage AND limiting voltage rise at the low voltage side in case of a short between two voltage levels.

2.2.5 Over-voltage and Electrostatic discharge [Challenges]

The following sources of over-voltages are to be considered:

- fuse interrupting over-current,
- load dump during crash situation,
- switching pulses of relays (if present (and not replaced by MOSFET's),
- inductive transients and
- electro static discharge pulses.

Applying the 58 V maximum 400 msec limit will impose very severe specifications on protection components or even on-chip protection systems. Silicon processes are in development giving breakdown voltages of 75 V, 80 V and even 90 V [20]. Moving to higher breakdown voltage gives reduced possibility for low R_{dson}. (high current switching) as $R_{dson} \propto V_{Breakdown}^2$. Also, there exists a higher risk for personal injury when assuming physical contact in damp and salty environments.

The inclusion of an (active) clamping function onto each chip however will enlarge the silicon area (and costs) therefore. This seems yet to be fully comprehended as the pulse protection requirements have not accurately been defined.

Table 4. Comparison of the 42V-PowerNet with today's 14 V system

	Nominal voltage	Max. clamping voltage	Min. semiconductor breakdown voltage
42 V	42 V	58 V	75 V or 90 V
14 V	14 V	40 V	60 V

The challenge arises in constructing components to clamp the occurring over-voltage to such a level that it is below the breakdown voltage of the electronics of the remainder of the system. The lower the clamping voltage the better for the electronics, however to maintain the same clamping characteristic the device will need to be larger. Components may either be single or array-type of devices.

2.2.6 Environmental issues, lead free and halogen free soldering [Challenges]

Industry is striving towards environmentally friendly solutions; a "black" list of forbidden substances exists and is being updated regularly, as well as a "grey" list of substances to be abandoned in future. The "grey" list incorporates soldering of Printed Circuit Board assemblies using Pb- and Halogen free solders. At this point in time, the (long term) reliability of the new solder types has not been sufficiently proven.

The challenge arises when providing an environmental solution without impairing the long-term functionality and reliability of the PCB-assembly or loosing backward compatibility.

2.3 Potential Risks [Challenges]

Potential risks of a 42 V system using incorrectly specified circuit protection components, are:

- under or over specification of components, resulting into non optimised system and high penalty,
- replacement of over-current/over-voltage protection component with one not being 58 V rated, resulting into a non-protected system,
- cable harness not protected resulting into vehicle fire,
- infotainment electronics not protected resulting into expensive replacement for the end-customer and/or vehicle fire and
- electronic control modules malfunctioning resulting into unsafe driving conditions for the end-customer.

All of the above can result into Automotive OEM and supply chain liability. Reference is made to the National Health and Traffic Safety Administration (USA; see http://www.nhtsa.com) and to the upcoming awareness in Europe of consumer organisations, currently active in increasing for to-days 12 V vehicles:

- the durability requirements during driving conditions and
- the crash test requirements.

Fig. 9. Crash test video example obtained from consumer organisations 1999

3 Most realistic assumptions

3.1.1 Over-current [Assumptions]

Although full details of the 42V-PowerNet system are still not finalised and differ from AOEM to AOEM, some (draft) standards for fuses have been issued:

- DIN 72581 part 3 (March 2001) Road vehicles – Blade type fuse links
- DIN 72581 part 4 (November 2001) Road vehicles – User guide
- SAE J 2576 (approved Draft Jan 2002) Blade fuses – 42 V System

These standards assume short circuit test to be equal to less severe as 60 V/1000 A/2.5 msec, being the best guess possible at this time and awaiting better information from the market.

Littelfuse have adopted above best guess in order to allow some latitude for different battery-technologies to come and component aging influences; this although:

- The lead-acid system with 12 kW starter-alternator may be sufficiently specified with:
  1. 58 V/200 A/2.5 msec, where 58 V is the dynamic over-voltage
  2. 52 V/250 A/2.5 msec, where 52 V is the static over-voltage
  3. 43 V/1000 A/2.5 msec, where 43 V is the max. voltage for engine running.
- The best guess requirements may not be stringent enough for systems using 58 V static and dynamic over-voltage combined with battery internal resistance lower than 58 V/1000 A = 58 mOhm.

3.1.2 Over-voltage [Assumptions]

The best assumption for the 42 V over-voltage requirements are anticipated as follows:

- Harness resistance will change: while in theory it could increase up to 9 times, in practice it is unlikely to increase by more than perhaps 4 to 6 times [5].
- The standard impulse waveforms (excluding Load Dump) as already specified will remain the same. [Paragraph 2.1.2]

- the 58 V, 400 ms, Maximum Dynamic Voltage limit while nominally applying to the entire system will in practice only be used to verify that modules are sufficiently designed to withstand the suppressed load dump pulse from the alternator.
- Locally generated transients suppression levels will have to be above 58 V for several reasons:
    1. Avoidance of each module playing the role of central suppressor
    2. Local high voltage bus system requirements e.g. 80 V direct fuel injection.

As per ISO/DIS 7637 Parts 1 and 2 “Road vehicles – Electrical interference by conduction and coupling – Part 2: Electrical transient conduction along supply lines only on vehicles with nominal 12 V and 24 V supply voltage“ currently under evaluation for classic 12 V/24 V systems:

- Suppression components and suppression levels used by different car manufacturers are not standard and
- Values for classification of functional status, test pulse severity, and presentation of results are to be agreed between vehicle manufacturer and equipment manufacturer.

3.1.3 Electro Static Discharge [Assumptions]

The need for ESD protection will not change – the threats will remain or even increase with smaller electronic component size, and quiescent operating voltage increase.

3.2 Proposed solutions.

All of the 42 V solutions below are to be considered draft solutions on so-called best guess specifications. This industry has not yet established full set of requirements.

3.2.1 Over-current [Solutions]

During latest SAE Motor show in Detroit, Littelfuse have shown Product Bulletin on 42 V solutions, which includes for MINI42, JCASE42 and MAXI42 dimensional and performance characteristic description of product as well as dimensions on proposed mating cavities for prevention of insertion of non 58 V compliant product. Further prototype parts exist for ATO42, MIDI42, MEGA42, and CablePro; this awaiting final technically justified market requirements.

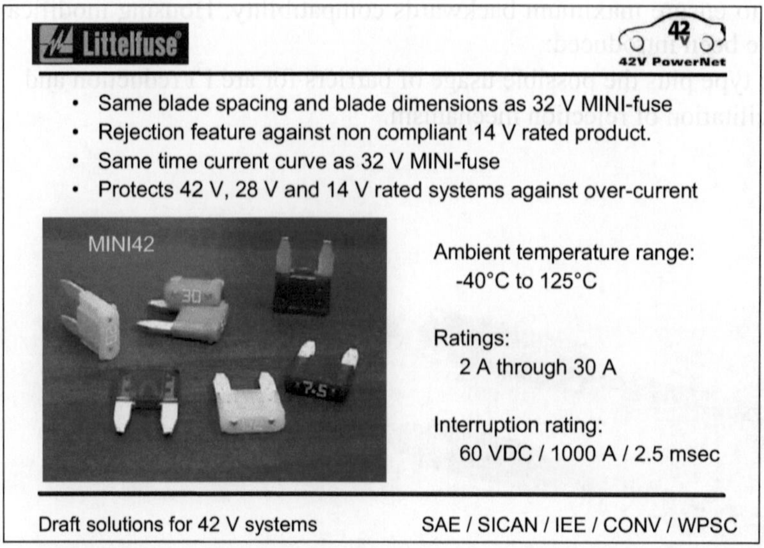

Fig. 10. MINI-42

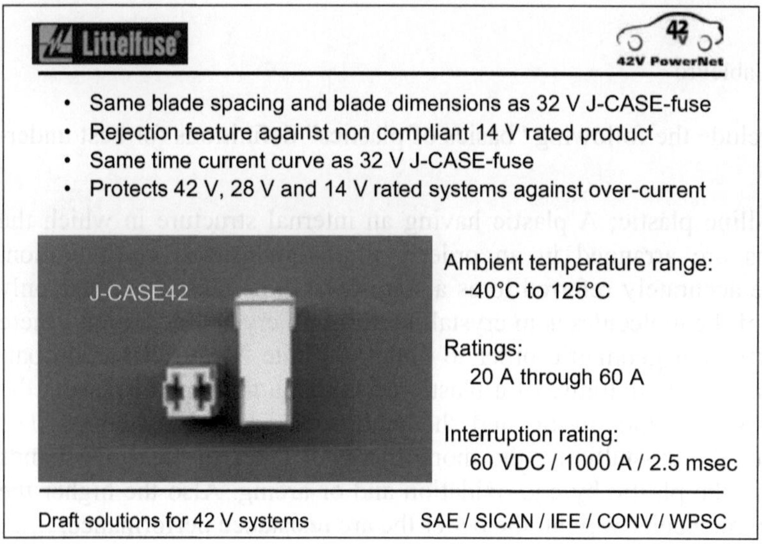

Fig. 11. JCASE-42

Fuse elements have proven to be sufficiently robust in design and functionality to ensure maximum backwards compatibility. Housing modifications have been introduced:

- plastic type plus the possible usage of barriers for arc I^2t reduction and
- the facilitation of rejection mechanism.

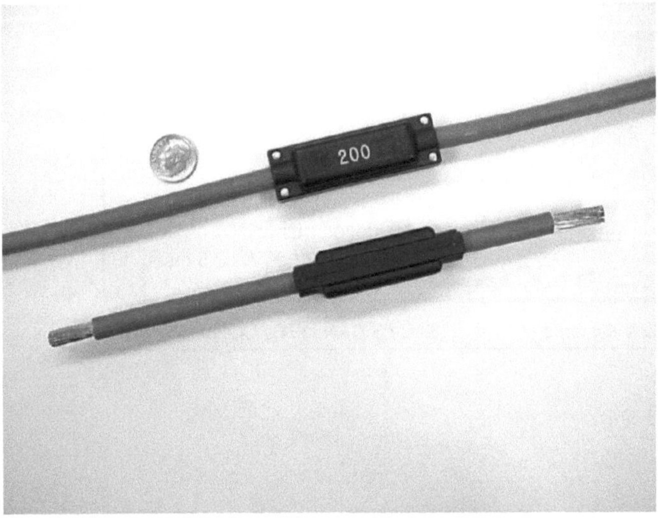

Fig. 12. CablePro

We include the following "basics of plastics" definitions for best understanding:

- Crystalline plastic: A plastic having an internal structure in which the atoms are arranged in an orderly three-dimensional configuration. More accurately referred to as a semi-crystalline plastic because only part of the molecules is in crystalline form. In crystalline regions, there is sufficient geometric order to obtain definite X-ray diffraction patterns. The crystallinity of a plastic depends on the type of plastic, the additives to the plastic and the parameters during moulding. The higher the crystallinity, the more energy it takes to break the bonds within the plastic by e.g. oxidation and or arcing. Also the higher the chain molecular weight, the better the arc resistance performance.
- Amorphous plastic: A plastic that has no crystalline component, no known order or pattern of molecule distribution and no sharp melting point.

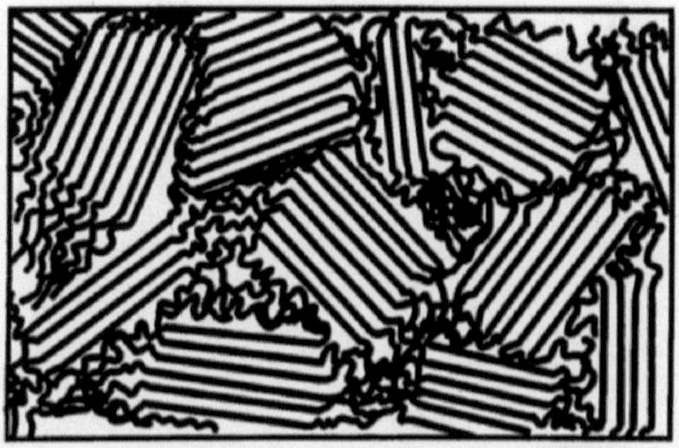

Fig. 13. Crystalline and amorphous regions inside a piece of plastic

- Cross-linking: The setting up of chemical links between the molecular chains. When extensive, the cross linking makes an infusible super-molecule of all the chains.
- Annealing: In plastics, heating to a temperature at which the molecules have significant mobility, permitting them to re-orient to a configuration having less residual stress. In semi-crystalline plastics, heating to a temperature at which retarded crystallisation or re-crystallisation can occur.
- An-isotropic: Exhibiting different properties when tested along axes in different directions.

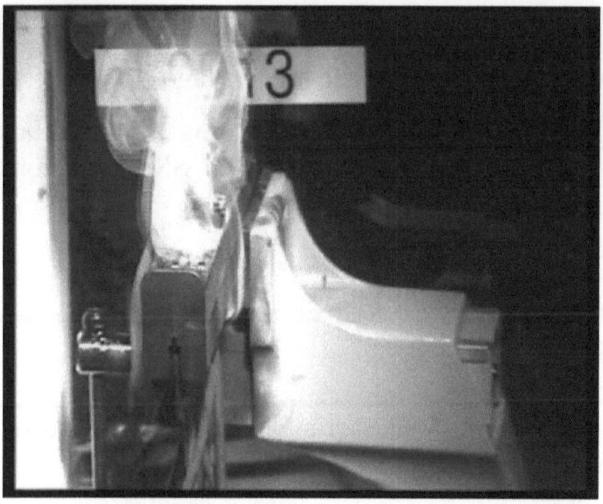

Fig. 14. IABG testing of 42 V fuses using UL94-HB fuse block materials

Note that as, in the absence of an arc-fault protector, the fuse-holder or fuse-block may be part of an arc at the location of a connector, the above plastic considerations are not only to be held for fuse-housing, but also for the fuse holder and also for cable insulation, which properties depend on ambient temperature, mechanical and electrical loading as well as on environmental influences.

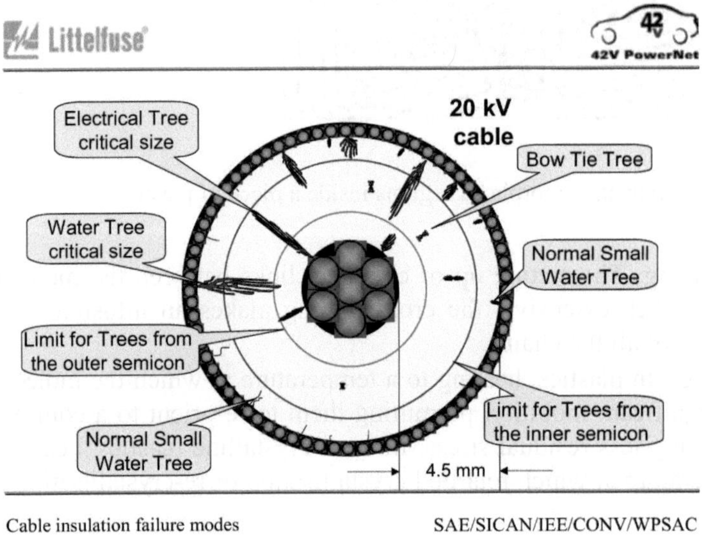

Fig. 15. Cable failure mode example

3.2.2 Over-voltage and Electro-static Discharge [Solutions]

Signal and Control Electronics Lines at classic bus voltage

For the protection for signal/control electronics operating at lower bus voltages, the present ML, MLE, MLN, AUML series for SMD solutions and the ZA series for through-hole solutions:

- 1812- and 1210-sized varistors for switching transient suppression,
- 0805- 0603- and 0402-sized varistors for ESD suppression and
- 7 – 20 mm leaded devices, 18 – 36 V.

will be equally valid.

Alternative strategies may be use of SP7 series diode arrays or Pulseguard™ devices for ESD protection.

Signal and Control Electronics Lines at 42 V bus voltage

For the protection for signal/control electronics operating at 42 V bus voltages, one may choose as starting point:

- V42MLA1206 multi-layer varistor. Maximum clamping voltage is about 86 V. Minimum semiconductor breakdown voltage would need to be 90 V.
- V56CH8 multi-layer varistor. Maximum clamping voltage is about 107 V. Minimum semiconductor breakdown voltage would need to be 110 V.
- V56ZA range, 7 – 20 mm leaded varistor with Maximum clamping voltage is about 110 V. Minimum semiconductor breakdown voltage would need to be above 110 V.

Alternative strategies may be evaluated including the possibilities of SGT surgectors.

42V Power lines

As the 42 V forums to date have only specified a pulse for load dump and assumed the starter-alternator to be equipped with electronics limiting voltage to maximum of 58 V DC. This is a major challenge to two main types of over-voltage suppression, namely MOVs and Zener diodes. New types of suppressors with very low dynamic impedance and low temperature coefficient of clamping voltage will be required.

While Zener diodes have superior clamping performance to MOV's, the impact of especially temperature coefficient of clamping will make it difficult for this technology to meet a 48 V operating voltage with a 58 V maximum under transient conditions.

The problem in choosing an MOV, for example to meet the proposed maximum dynamic voltage limit is clearly illustrated by the fact that an 48V-rated Electrical Industries Association (EIA) 2220-sized surface-mount varistor presently clamps a 100 V (30 Ω) switching transient at 78 V and a 300 V (30 Ω) switching transient at 85 V. Consideration of this type of suppression device for power electronics protection will only be possible if the breakdown voltage of the silicon is raised to 90 V or above.

New developments in varistor technology include "Grain size control" technology currently under development will enable the production of higher rated voltage varistors with much improved clamping characteristics:

- Varistors are sintered electro-ceramic devices composed of Zinc oxide (ZnO) and other metallic oxides. The electrical properties are derived principally from the boundaries between the fired grains. Each boundary behaves as a zener diode and therefore the number of grains between electrodes determines the varistor "breakdown" voltage. EIA package dimension restrictions limit the number of layers that can be used – directly affecting the effective surface area of ZnO and hence the varistor V-I clamping characteristic. The number of electrodes within the structure determines the effective electrode area and therefore the voltage clamping performance. Increasing the rated voltage from say 14 V to 42 V means much larger electrode spacing and hence much larger overall multi-layer structured devices. To conform to EIA package dimensions the number of electrode layers would have to be restricted – directly affecting the effective surface area of ZnO and hence the varistor V-I clamping characteristic.

- "Grain size control" technology will produce varistor grains up to five-times smaller than today's, enabling larger number of layers and a consequent improvement in clamping characteristic to be achieved, see Fig. 16.

A secondary benefit of closer layers enables larger inner-electrode metalisation areas to used, further improving the effective varistor area. As can be seen from Fig. 16. there will be a considerable improvement in the clamping characteristic of the Multi-layer varistor over the range of currents of interest (3- to 10 Amperes) for automotive transient impulses, excluding load dump. Assuming silicon breakdown voltages of 75 V or 80 V, the Grain size control technology varistor becomes a must.

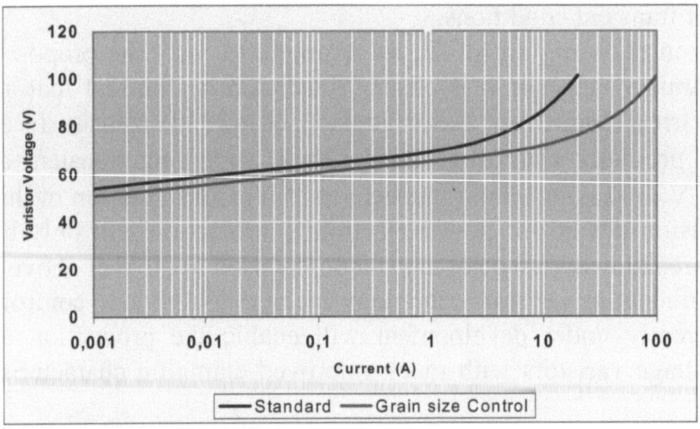

Fig. 16. Comparison of "Grain size control" technology 2220-size varistor clamping characteristic with to-days 42 V rated varistor

3.2.3 Arc fault protector

Current measurement together with Frequency analysis could be used to distinguish between normal currents and arc fault currents. Note this only applies for frequencies outside the vehicle's end-user normal switching behaviour. Usage of PWM driven 12 V end-users may complicate arc fault detection by frequency analysis and require other methodology to be developed.

Current measurement

To allow current measurements to be taken, LF have developed shunts in the range of 0.3 – 2.5 mOhm. The exact component value to be chosen depends on requirements for continuous current load at elevated ambient (LF have generated a wide range of SPICE models), magnitude of voltage signal (the smaller, the more prone it is for measurement error) and thermal EMF versus copper.

Fig. 17. MIDI sized shunt for current measurement

Protection of electrical system not involved in principal driving.

Taking the signals from airbag and in vehicle current measurement, a decision needs to follow that either a car crash or arc fault has occurred. Switching off all non-critical electronics applications, as e.g. all infotainment (navigation systems, access to internet etc.), may be done by e.g. activating MOSFET or thyristor to ground, which is connected in parallel to the main fuse. From the battery a large short circuit current then can flow to ground opening the main fuse. During short circuit flow, the MOSFET/thyristor may experience some damage as long as the fuse opens fast and reliably. Fuse and MOSFET/thyristor need replacement after the crash or arc fault occurrence; both need treatment from a specialist.

Reason to connect the electronic switch in parallel rather than in series is the fact that MOSFET leakage current rises with temperature and age (time). Being only once energised, the temperature of the MOSFET/thyristor is as low as possible, and hence the reliability as high as possible.

Protection of electrical systems involved in principal driving

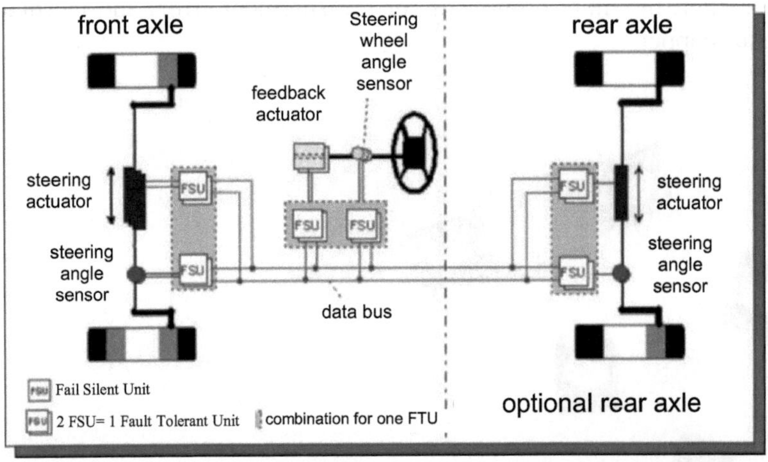

Fig. 18. Example of „Drive by Wire“

Principal vehicle functions such as electric braking, electric steering, active suspension, all need separate treatment as one does not want to loose the braking power, steering power or active suspension, if something would be wrong in one of the other electrical systems. This means:

- Braking, steering and suspension all would need their own arc fault protector, which can be triggered by arc fault signal of the particular circuit.
- If for each of the above same irreversible approach would be chosen being a main fuse blown by switching parallel MOSFET/Thyristor, this

would end up in vehicles not being able to brake, steer or have active suspension after an arc fault.

The main question becomes now how reliable the electrical systems can be made:

- If arc faults only occur during crash, then one could accept an irreversible device.
- If arc faults occur during normal use, one needs an indication that the vehicle is not ready to drive (again), needs transportation to a repair station, an expert, detailed investigation of the vehicle and perhaps a reversible device.

The considerations made above are related to any electrical (i.e. over-current and over-voltage) fault conditions:

- Both electronic components and applications can be quite complex. Design and Process Failure Mode and Effect Analysis may include many steps.
-

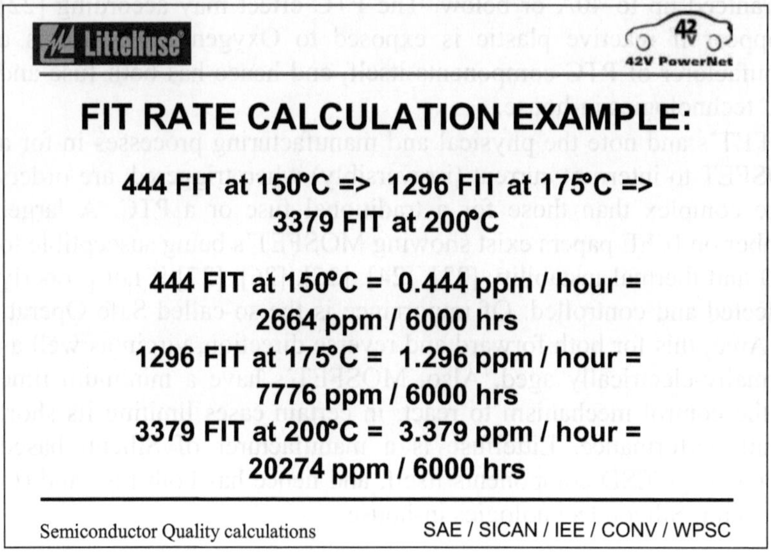

Fig. 19. Example of „ppm" calculation

After a smaller vehicle crash, it is practically impossible to determine if a piece of electronics has not been damaged electrically. Hence one may choose for over-current and over-voltage protection of all Electronic Control Unit input and output lines.

- Today's MOSFETS's being used electronic power switching are only to a mini-mum level protected against electrostatic discharge (ESD).
- No or limited history exists on use of large scale use of printed circuit boards with Pb-free and/or halogen free soldering.

3.3 Alternative strategies.

3.3.1 Over-current [Alternatives]

As over-current protection alternatives, we mention:

- PTC.'s [21] and note the physical processes in a PTC's to interrupt current (reversibly) are based on plastics with a number of additives. Their functionality is considered far more complex than compared to a fuse element (in which the metal fuse element melts and evaporates). Hence the tolerances achieved with PTC's are larger than demonstrated with traditional fuses. PTC short circuit rating is normally guaranteed up to 40A or below. The PTC effect may according [22] disappear if reactive plastic is exposed to Oxygen. Littelfuse is a manufacturer of PTC components itself, and hence has both fuse and PTC technologies in-house.
- MOSFET's and note the physical and manufacturing processes in/for a MOSFET to interrupt current (irreversibly) when triggered, are orders more complex than those for a traditional fuse or a PTC. A larger number on IEEE papers exist showing MOSFET's being susceptible to ESD and thermal instability [23], [24], [25], [26], [27] if not properly protected and controlled. Of importance is the so-called Safe Operating Area; this for both forward and reverse direction, virgin as well as thermally-electrically aged. Also, MOSFET's have a minimum time for the control mechanism to react, in certain cases limiting its short circuit performance. Littelfuse is a manufacturer of Silicon based Over-voltage/ESD components itself, and hence has both fuse and (to an extend) Silicon technologies in-house.

3.3.2 Over-voltage [Alternatives]

Assuming silicon breakdown voltages of 60 V and that the 58 V dynamic maximum voltage applies for the entire system and all transients, then it is conceivable that TVS is only possible with a device such as an internally-clamped Power MOSFET aided by an active control to detect and divert a transient.

Such components and control systems would be expensive in comparison to the existing passive component solutions. It is also possible that such devices would be required on all modules throughout the vehicle.

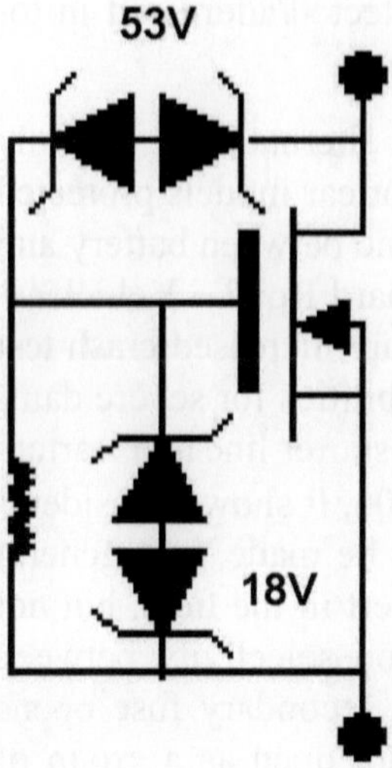

Fig. 20. Internally clamped power MOSFET

3.4 Modelling support: Why?

Reasons to support experimental investigations with theoretical modelling are:

- Minimising of system cost,
- Minimising of time to market,
- Optimising functionality,
- Optimising customer satisfaction and
- Minimising warranty claims.

For fuses thermal-electrical models have been reported by e.g. [28]. Littelfuse have a full library of thermal-electrical models for all ratings Automotive fuse series being ATO, MINI, JCASE, MAXI, MIDI, MEGA. This both for 32 V as for 58 V rated components.

For MOV's thermal-electrical models have been reported by e.g. [29]. Littelfuse have a full set of electrical models and are working on upgrading to include thermal behaviour.

3.5 Are today's systems fully protected?

3.5.1 Over-current [Protection]

Below two typical examples of items not fully protected/addressed in todays vehicles:

- The lines between battery and starter, battery and alternator are the only unprotected cables left, with a few OEM's and/or car models protected against over-current. The same applies for the line between battery and starter. With the introduction of ISO 3560 standard Road – Vehicles – Frontal fixed barrier or pole impact test procedure increased crash test requirements are to come and hence more possibilities for severe damage to battery-alternator-line, alternator, battery-starter line and starter.
- User-guide DIN 72581 part 4 was issued Nov. 2001. It shows considerations for the user of fuses (AOEM, 1rst tier) to be made, in a general manner. Requirements are "obvious" to the expert in the field, but not that obvious for the "non trained". We mention selectivity between main and secondary fuses in particular. If the secondary fuse opens due to an over-current, the main fuse should not open as a group of end-users may become un-energised (unsafe during driving).

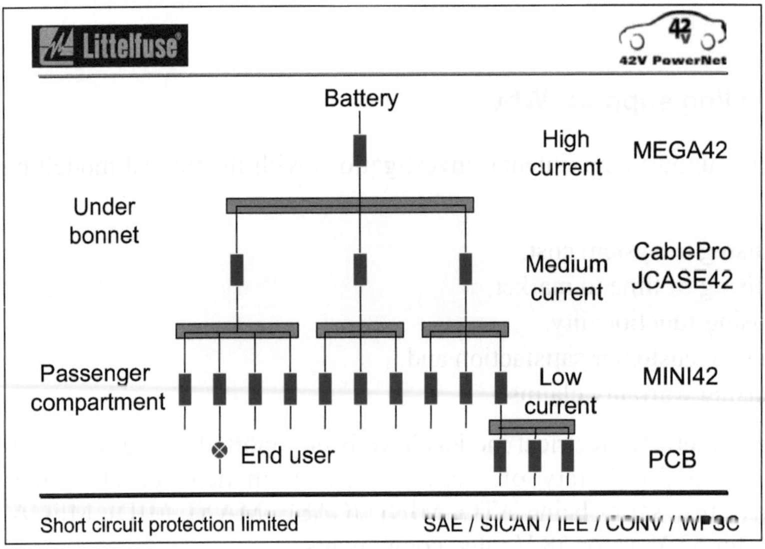

Fig. 21. Selectivity definition

To support customers in the selection of over-current protection components, Littelfuse have established User-guide [32] in excess of DIN 72581 part 3 and Draft ISO 8820 part 2 documents. The User-guide includes spreadsheet type of calculation and makes reference to the Littelfuse SPICE simulation tools.

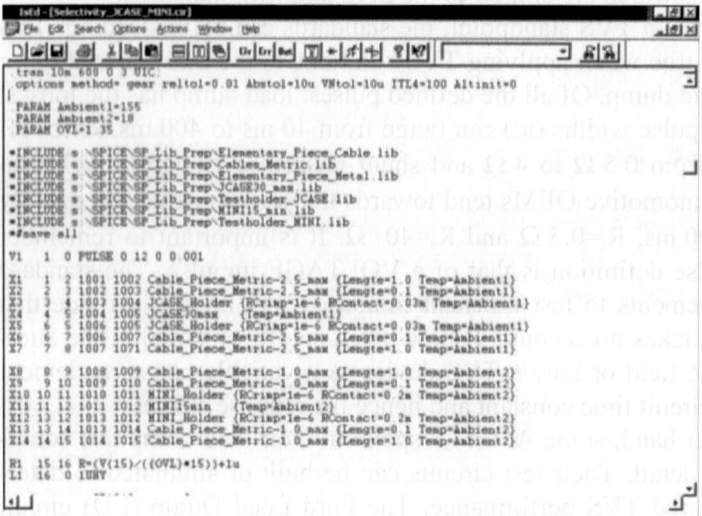

Fig. 22. Selectivity between main and secondary fuses – SPICE code.example

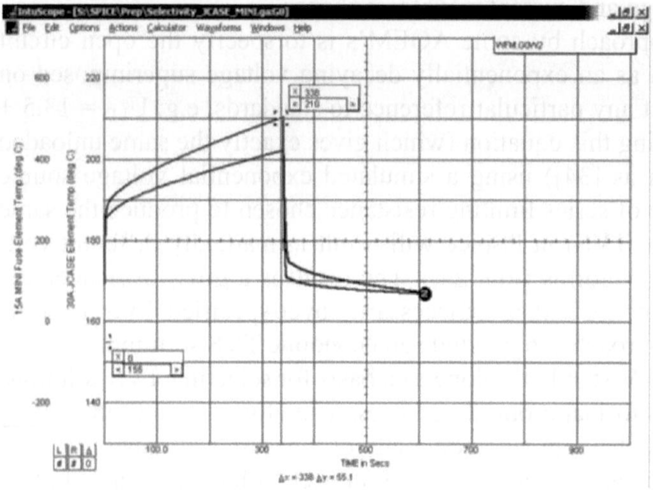

Fig. 23. Selectivity between main and secondary fuses - output example

3.5.2 Over-voltage and Electro Static Discharge [Protection]

In today's automotive environment there are many standards covering the over-voltage transient threats. These standards, e.g. SAE J 1113, ISO 7637, describe in detail the transient impulses for compliance testing of automotive modules. In addition many AOEM's have their own in-house standards which are similar to the SAE and ISO standards.

However from a TVS standpoint, the standards can be misleading. This is particularly true when applying TVS solutions to the most crucial pulse i.e, pulse 5-load dump. Of all the defined pulses, load dump has the loosest specification: pulse widths (td) can range from 40 ms to 400 ms, series resistance (R_i) from 0.5 Ω to 4 Ω and shunt resistance (Rs) from 0.7 Ω to 40 Ω. Most Automotive OEMs tend towards the worst case and frequently specify: t_d=400 ms, R_i=0.5 Ω and R_s=40 Ω. It is important to remember that the impulse definition is that of a VOLTAGE impulse. The standard defines requirements to test transient immunity, not how to provide that immunity and takes no account of how much energy is stored in the alternator magnetic field or how a TVS device as a variable resistive element modifies the circuit time constant and hence the impulse shape.

On the other hand, some AOEM's spell out their load dump test conditions in great detail. Their test circuits can be built or simulated to determine the required TVS performance. The Ford Load Dump (LD) circuit [34], for example, contains a shunt resistance of 0.7 Ω which considerably modifies the stress seen by the TVS. Fig. 24. shows the varistor clamping voltage and current when simulated in the Ford load dump circuit. Notice the differing voltage and current waveform shapes.

An alternate approach by some AOEM's is to specify the open circuit voltage waveform as an exponentially decaying voltage superimposed on the battery without any particular reference to standards, e.g: $U_{LD} = 13.5 + 60e^{-t/0.13}$. Simulating this equation (which gives exactly the same unloaded voltage waveform as [34]) using a simulated exponential voltage source (and using a value of series limiting resistance chosen to produce the same peak current in the TVS) in PSpice will result in markedly different voltage and current waveforms (Fig. 24.). The result is a gross overstatement of the dissipated energy and an unnecessarily over-specified TVS. In some cases, the latter approach can preclude an economic TVS solution.

Using the 12/24 V standards alone as a basis for selecting TVS solutions can therefore lead to inappropriate choices. Therefore accurate modelling of the load dump behaviour (including any of the starter-alternator outputs due to failures such as sudden battery disconnect, voltage regulator failure or short between rotor and stator in case of crash) needs to be carried out and the results incorporated into any new standard.

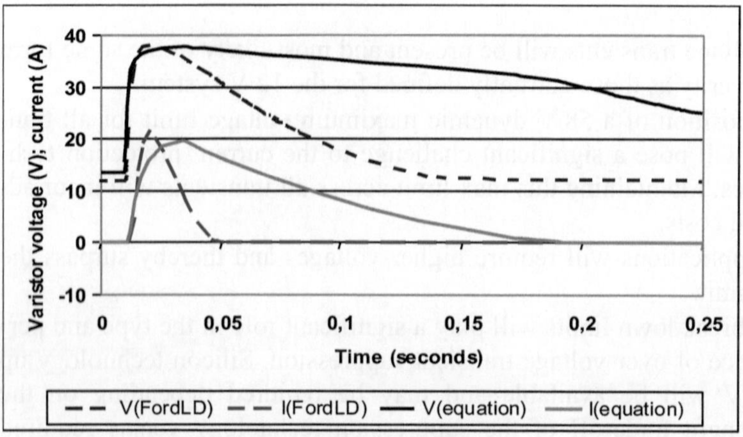

Fig. 24. Comparison of the simulation outputs of a Littelfuse V18AUMLA2220 varistor in the Ford load dump circuit with those achieved using simulated exponential voltage sources.

4 Conclusions

With respect to the classic voltage systems we notice an increased need for support in selecting correct circuit protection components. With respect to the development and introduction of a 42 V system we notice:

- Such a system is highly complex and has many significant implications. Hence it is of a different magnitude than when upgrading one classic voltage architecture into it's successor.
- A clear understanding of what is or will be required is still being developed.

Specifically on over-current and over-voltage for 42V-PowerNet, we summarise:

4.1 Over-current [Conclusions]

- Batteries, Starter-generator, UltraCap specifications and end-user components specifications have not finalised; therefore only draft short circuit parameters are available.
- Requirements are subject to change.

4.2 Over-voltage [Conclusions]

- Over-voltage transients will be present and most likely of the same form and severity as those currently defined for the 12 V system.
- The imposition of a 58 V dynamic maximum voltage limit for all transients will pose a significant challenge to the current protection technologies. Maintaining this max limit across all transients will incur additional costs.
- Some applications will require higher voltages and thereby surpass the 58 V limit.
- Silicon breakdown limits will play a significant role in the type and performance of over-voltage transient suppression. Silicon technology up to 90 V will be available and may be required depending on the cost/benefit trade-off of the suppression technology versus required performance.
- Suppression component manufacturers are currently working on next generation technologies to meet the expected tighter performance requirements.
- Requirements are subject to change.

Littelfuse is the world leader and first in circuit protection devices. We generate solutions for all major and electrical and electronic market world wide. We pride ourselves in being best in class for tight tolerances, innovative components in over-current, over-voltage and ESD.

5 References

[1] Matouka M, Design considerations for higher voltage automotive systems, SAE Paper 911654, Warrendale PA, USA, 1991

[2] Davis B; Frank R, Valentine R, The Impact of Higher System Voltage on Automotive Semiconductors, SAE Paper 911658, Warrendale PA, USA, 1991

[3] Eisenacher, Sicks, Bordnetzspannung 42 V, AA-I-3 Schriftstück Nr. 01/98, Frankfurt, Germany, 1998

[4] Convergence 2000 Proceedings, SAE P-360, Detroit MI, USA, 2000

[5] Miller J, Goel. D, Kaminski D, Schöner H, Jahns T, Making the case for a Next generation Automotive Electrical System., http://auto.mit.edu, Website Massachusetts Institute of Technology

[6] http://www.sci-worx.com, Website SCI-WORX

[7] Eisenacher K, Cheynet J, Road Vehicles - Electrical and electronic equipment for 42 V network (PowerNet), ISO TC 22 N2182, new work item proposal, incl. WD 42V-1E and –2E, Detroit MI, USA, 2000

[8] Jaspar J, Brown W, Oh S, Travis W, Wally P, Fuses for Future Vehicles with 42 V rated Electrical Systems, SAE Paper 2000-01-0137, Detroit MI, USA, 2000
[9] Hetzmannseder E, Zuercher J, 42 V DC Arc Faults – Physics and Test Methods, Eaton Innovation Center
[10]Sakiyama K, Ide T, Akashi K, A new connector for 42 V Automotive electrical System, SAE Paper 2001-01-0724, Detroit MI, USA, 2001
[11]Henry R, Lequesne B, Chen S, Ronning J, Xue Y, Belt driven Starter-Generator for future 42 V Systems, SAE Paper 2001-01-0728, Detroit MI, USA, 2001
[12]Kusumi H, Yagi K, Ny Y, Abo S, TOYOTA Motor Corporation
Sato H, Furuta S, Morikawa M, DENSO Corporation.
42 V Power Control System for Mild Hybrid Vehicle (MHV), SAE Paper 2002-01-0519, Detroit MI, USA, 2002
[13]Krappel A, et. al., Kurbelwellenstartergenerator (KSG) – Basis für zukünftige Fahrzeugkonzepte, ISBN 3-8169-1808-5, Expert Verlag, Renningen-Malmsheim, Germany, 1999
[14]Inspira Battery manual for 12 V and 36 V Batteries, Rev. H.
Johnson Controls, Inc; Automotive Systems Group, Battery Division Milwaukee, WI, 2000
[15]Guttowski S, Thomas J, Hurley W, Kassakian J, The Impact of Transition from 14 V to 42 V on EM Noise of Automotive Relays
[16]Le Bars P, Regini A, 42 V Load Dump and centralised active suppression, IEE Paper 930012, London, UK, 2000
[17]Alternator/Regulator Design for Automotive Charging Systems, MD 0156, Analogy Inc., Beverton, OR, 2000
[18]Caliskan V, Modelling and Simulation of a Claw Pole Alternator: Detailed and Averaged Models, MIT Laboratory for Electromagnetic and Electronic Systems, 1998
[19]Champlin K, Bertness K, DFRA/FDIS – A New Technology for Battery Instrumentation and Testing, 2002
[20]Zitta H, Schmidt-Habich H, Stecher M, SPT4/90 A Voltage Derivative of the Smart Power Technology opens the Door to the 42V-PowerNet, Infineon Technologies at 42V-PowerNet: The first Solutions, Villach, Austria, Sept 1999
[21]Walsh M, Gaynier J, DeGrendel G, The Use of Polymeric PTC Devices in Automotive Wiring Systems, SAE Paper 930012, Detroit MI, USA, 1993
[22]Meyer J, Stability of Polymer Composites as Positive–Temperature-Coefficient Resistors, Polymer Engineering and Science, Vol. 14, No 10, Oct 1974, Texas Instruments Inc, Attleboro MA, USA
[23]Consoli A, Gennaro F, Testa A, Consentino. G, et. al., Thermal Instability of low voltage Power MOSFET's, IEEE Transactions on Power Electronics, Vol. 15 No 3, May 2000, Piscataway NJ, USA
[24]Sitte R, 3D Visualisation of Deep Submicormeter Transistor Characteristics, IEEE Transactions on Semiconductor Manufacturing, Vol. 15 No 1, May 2000, Piscataway NJ, USA

[25]Semenov O, Pradzynsi A, Sachdev M, Impact of Gate induced Drain Leakage on Overall leakage of submicrometer CMOS VLSI Circuits, IEEE Transactions on Semiconductor Manufacturing, Vol. 15 No 1, Feb 2002, Piscataway NJ, USA

[26]Winarski T, Dielectrics in MOS Devices, DRAM Capacitors, and Inter-Metal Isolation, IEEE Electrical Insulation Magazine, Vol. 17 No 6, Nov-Dec 2001, Piscataway NJ, USA

[27]Oldervoll F, High Electric Stress and Insulation Challenges in Integrated Micro-electronic Circuits, IEEE Electrical Insulation Magazine, Vol. 18 No 1, Jan-Feb 2002, Piscataway NJ, USA

[28]Herel van J, Thermal modelling of a miniature fuse in PSPICE (In Dutch), Technical University Eindhoven, Elektrotechniek, Elektrische Energiesystemen, Report EG/92/610, Eindhoven, The Netherlands, 1992

[29]Lat M, Analytical method for performance prediction of metal oxide surge arrestors, IEEE Trans. on Power Apparatus and Systems Vol. 104, No 10, Oct. 1985, Piscataway NJ, USA

[30]Martzloff F, Matching surge protective devices to their environment, IEEE Conference Industrial Applications Society, 1983, Piscataway NJ, USA

[31]Densley J, Aging mechanisms and Diagnostics for Power Cables – An overview, IEEE Electrical Insulation Magazine, Vol. 18 No 1, Jan-Feb 2002, Piscataway NJ, USA

[32]Jaspar J, User-guide for classic 12 V/24 V Rated Systems Rev. B, Part 1, Over-current, LITTELFUSE BV, Feb 2002, Utrecht, The Netherlands

[33]McLoughlin N, Jaspar J, User-guide for classic 12 V/24 V Rated Systems, Rev. A, Part 2, Over-voltage; document under preparation, LITTELFUSE BV, 2002, Utrecht, The Netherlands

[34]Ford Engineering Specification, ES-XW7T-1A278-AB, Part 3, section CI240

[35]Demeis R, Engineers transition to 42 V, Global Design News, Oct. 2001

Contacts

- LITTELFUSE BV.
  E-mail: jjaspar@littelfuse.nl
  Proost Wetering 81, PO Box 2023
  3500 GA Utrecht, The Netherlands.
- LITTELFUSE LTD.
  E-mail: nmclough@littelfuse.com
  Ecco Road
  Dundalk, Ireland

Use of PolySwitch PPTC Protection in Automotive Applications

Werner Gretzke

Tyco Electronics Raychem GmbH, Ottobrunn

Introduction

PPTC device technology has been widely applied to overcurrent and overtemperature circuit protection designs in portable electronics, cell phones, computers, and telecommunications equipment. New Automotive Electronics Council standards for passive components, including PPTC devices, are advancing the acceptance of this technology in the automotive industry.

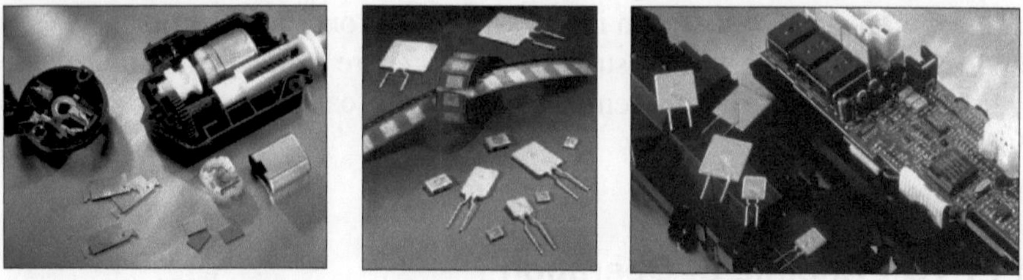

Fig. 1. PPTC devices

New vehicle designs rely heavily on electronic circuits and motorized accessories – such as power windows, power seats, sunroof controls, and telematics – to reduce cost, improve reliability, and add functionality. The current move from 14 V systems to 42 V systems is a direct response to the demand for fuel efficiency and even higher power requirements associated with the increasing electrical and electronic systems in the automobile platform.

Because the electrical system represents a large percentage of a vehicle's cost and weight, it requires adequate protection against short circuits and overloads. Current limiting can be accomplished by using a resistor, fuse, switch or PTC (positive temperature coefficient) device. Resistors are rarely an acceptable solution because an expensive high-power resistor is usually required. One-shot fuses may be used, but they may fatigue, and must be replaced after a fault event. The limitations of bimetallic switches include cycling and the potential for contacts to weld shut. CPTC (ceramic positive temperature coefficient) devices tend to have high resistance and power dissipation characteristics. They are also relatively large and may be vulnerable to cracking as a result of shock or vibration.

In many automotive applications the preferred solution is the PPTC (polymeric positive temperature coefficient) device, which has low resistance in normal operation and high resistance when exposed to a fault. Resettable PPTC devices help prevent overcurrent and overtemperature damage to automotive electrical equipment, power distribution systems, signal distribution systems, or electronic components that may result from an electrical short or electrically overloaded circuits.

Like traditional fuses, PPTC devices limit the flow of dangerously high current during fault conditions. Unlike traditional fuses, PPTC devices reset after the fault is cleared and power to the circuit is removed. Because the PPTC device does not usually have to be replaced after it trips, and is small enough to be mounted directly into the motor or on the circuit board, it can be located inside electronic modules, junction boxes, and power distribution centers. This design architecture allows for placement of electronic modules and systems in inaccessible locations and permits the use of smaller wires which can result in smaller wire harnesses and a cable weight reduction that some engineers estimate can be as much as 40 to 50%.

1 PPTC Principle of Operation

PPTC circuit protection devices are made from a composite of semi-crystalline polymer and conductive particles. At normal temperature, the conductive particles form low-resistance networks in the polymer. However, if the temperature rises above the device's switching temperature (T_{Sw}) either from high current through the part or from an increase in the ambient temperature, the crystallites in the polymer melt and become amorphous. The increase in volume during melting of the crystalline phase causes separation of the conductive particles and results in a large non-linear increase in the resistance of the device.

The resistance typically increases by three or more orders of magnitude, as shown below. This increased resistance protects the equipment in the circuit by reducing the amount of current that can flow under the fault condition to a low, steady state level. The device will remain in its latched (high resistance) position until the fault is cleared and power to the circuit is removed – at which time the conductive composite cools and re-crystallizes, restoring the PPTC to a low resistance state and the circuit and the affected equipment to normal operating conditions.

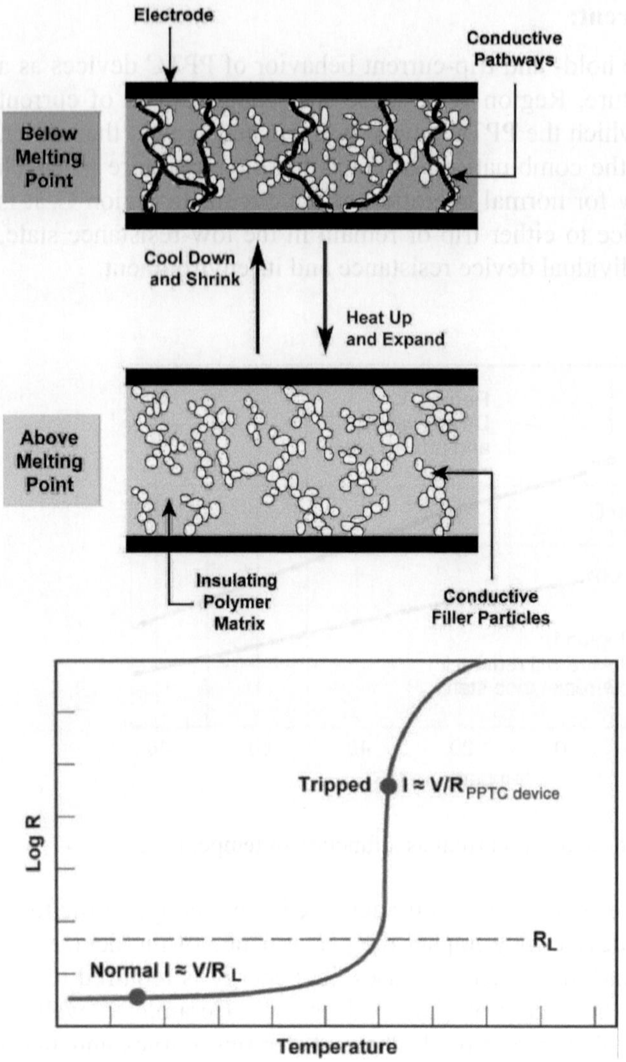

Fig. 2. PPTC devices protect the circuit by going from a low-resistance state to a high-resistance state in response to an overcurrent or overtemperature condition.

2 Design Considerations for PPTC Devices

Some of the critical parameters to consider when designing PPTC devices into a circuit include device hold- and trip-current, the effect of ambient conditions on device performance, device reset time, leakage current in the tripped state, and automatic or manual reset conditions.

Hold and Trip Current:

Fig. 3. illustrates the hold- and trip-current behavior of PPTC devices as a function of temperature. Region A describes the combinations of current and temperature at which the PPTC device will trip and protect the circuit. Region B describes the combinations of current and temperature in which the device will allow for normal operation of the circuit. In region C, it is possible for the device to either trip or remain in the low-resistance state, depending on the individual device resistance and its environment.

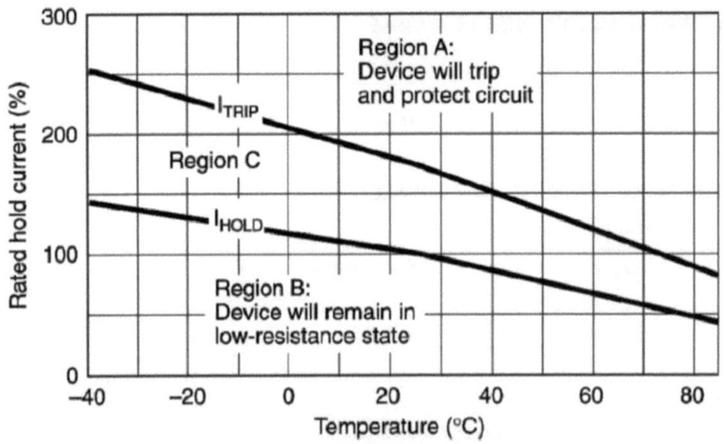

Fig. 3. Example of hold- and trip-current as a function of temperature.

Since PPTC devices can be thermally activated, any change in the temperature around the device may impact the performance of the device. As the temperature around the device increases, less energy is required to trip the device and thus hold current (I_{HOLD}) decreases. Ceramic as well as polymeric PTC manufacturers provide thermal derating curves and I_{HOLD} versus temperature tables to help the designer select the appropriate device.

Effect of Ambient Conditions on Device Performance:

The heat transfer environment of the device can significantly impact device performance. In general, by increasing the heat transfer of the device there will be a corresponding increase in power dissipation, time-to-trip and hold-current. The opposite will occur if the heat transfer from the device is decreased. Furthermore, changing the thermal mass around the device will change the time-to-trip of the device.

The time-to-trip of a PPTC device is defined as the time needed, from the onset of a fault current, to trip the device. Trip time depends upon the size of the fault current and the ambient temperature.

A trip event is caused when the rate of heat lost to the environment is less than the rate of heat generated. If the heat generated is greater than the heat lost, the device will increase in temperature. The rate of temperature rise and the total energy required to make the device trip depends on the fault current and heat transfer environment.

Under normal operating conditions, the heat generated by the device and the heat lost by the device to the environment are in balance.

$$I^2R = U(T-T_A) \qquad (4.1)$$

Where:

I = current flowing through the device
R = resistance of the device
U = overall heat-transfer coefficient
T = temperature of the device
T_A = ambient temperature

Increases in either current, ambient temperature, or both will cause the device to reach a temperature at which the resistance rapidly increases. This large change in resistance causes a corresponding decrease in the current flowing in the circuit, protecting the circuit from damage.

The hold current is the highest steady-state current that a device will hold for an indefinite period of time without transitioning from the low- to the high-resistance state. Hold current can be fairly accurately defined by the heat transfer environment and can be impacted by a multitude of design choices, such as:

- Placing the device in proximity to a heat-generating source such as a power FET, resistor, or transformer, resulting in reduced hold current, power dissipation and time-to-trip,

- Increasing the size of the traces or leads which are in electrical contact with the device, resulting in increased heat transfer and larger hold current, slower time-to-trip and higher power dissipation or
- Attaching the device to a long pair of wires before connecting to the circuit board, increasing the lead length of the device, resulting in a reduction of the heat transfer and lowering of the device's hold current, power dissipation, and time-to-trip.

The PPTC devices' low resistance, fast time-to-trip, and low profile help improve electronics reliability in a small footprint. They are compatible with high volume electronics assembly techniques, and are available in surface-mount, radial-leaded, or custom configurations, with a wide range of voltage, current, resistance and temperature specifications. To select the best device for a specific application, the circuit designer should consider the following design criteria.

1. *Choose the appropriate form factor:* Select from radial-leaded, surface-mount, or chip parts. For mounting on circuit boards, radial-leaded or surface-mount is preferred. Radial-leaded parts are typically wave-soldered to the board. Surface-mount parts are typically reflow-soldered to the board. Chip parts are designed to be held in clips, usually in an electric motor. These parts are often custom designed for specific applications.
2. *Choose a voltage rating:* A PPTC device should be chosen with a voltage rating that equals or exceeds the source voltage in a particular circuit. Also, the expected fault voltage should not be greater than the PPTC voltage rating. When a PPTC device trips, the majority of the circuit voltage will appear across it, because it will be the highest resistance element in the circuit. Parts with voltage ratings suitable for the 42V PowerNet are already available today, with more to follow.
3. *Choose a hold current rating (at the proper ambient operating temperature)*: Hold current is defined as the largest steady state current the PPTC device can carry without tripping into a high resistance state, at the specified ambient temperature. Because it is a thermal device, the hold current for a PPTC device decreases with increasing temperature. The actual value of the hold current for a given device and temperature may be obtained from the PPTC device manufacturer. The designer must choose a PPTC device with a hold current at the maximum ambient temperature equal to or greater than the steady state operating current.
4. *Check trip time*: PPTC device manufacturers can provide accurate time-to-trip curves illustrating how quickly the PPTC device trips at various currents. The designer should determine what fault currents may occur,

and how quickly the most sensitive system components may be damaged at these currents, and then select a PPTC device that trips before these components are damaged. In many applications, there is a start-up surge current from a capacitance or motor. Normally this in-rush does not contain enough energy to trip the PPTC device, but designers should confirm performance in their application over the range of expected ambient conditions.

5. *Check maximum interrupt current:* The PPTC normally has a maximum interrupt current rating, i.e., the maximum fault current that the device consistently interrupts while remaining functional.

3 Applications for Resettable Circuit Protection in Automotive Electronics

The transition to 42V-PowerNet and the interim dual-voltage network strategy offers many opportunities for innovation in the electrical and electronic system architecture. Decentralization of power distribution, more complex electronic modules, and smaller, localized wire harnesses are just a few areas of conversion in which resettable circuit protection can play a role.

Wire Harness Protection

Increasing power demands have resulted in complex wire harnesses that add wires, weight and packaging constraints to automobiles. Each electrical circuit requires adequate protection against short circuits and overloads, and although each load theoretically can be protected with its own dedicated fuse, fuses must be replaced when they blow. This characteristic requires that fuses be mounted in accessible fuse boxes—a requirement that dictates system architecture and forces packaging and system layout compromises.

The conventional solution groups similar circuits together and protects them all with a single fuse. The fuse must be sized to carry the sum of the currents drawn by each of the protected loads; and, in order to limit risk of damage and fire, the wires feeding from the fuse to each of the loads must be chosen according to the fuse size selected. This design practice often results in oversized wires with high current-carrying-capability feeding loads that require relatively low currents. Using heavy-gauge wire also requires use of larger terminals and connectors, which further increases cost, size, and weight. It also increases harness weight, and the weight of the automobile, which has an effect on fuel efficiency.

Because PPTC devices reset when a fault condition clears and power is removed from the circuit, they do not generally require routine replacement or service and can be placed inside doors, in switch assemblies, behind instrument panels, in electronic modules, and in other inaccessible areas within the vehicle. As shown in Fig. 4., the option of locating circuit protection devices strategically throughout the vehicle also allows for power to be routed via the most direct and efficient route, rather than through a central fuse box, which reduces the number of wires in the harness, and allows for a reduction in their weight and length.

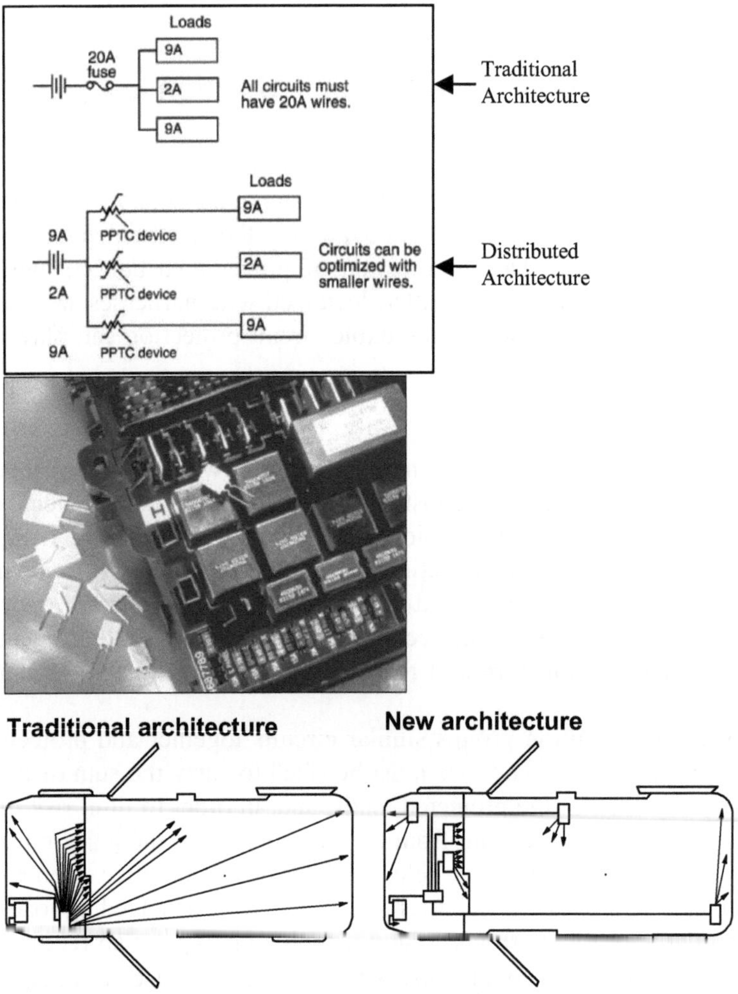

Fig. 4. PPTC devices can be used in distributed electronic system architectures to help reduce wire size.

Electronic Control Module Protection

As more and more circuitry is packed into smaller and smaller packages, the width of the copper traces on printed circuit boards (PCB) is reduced. Because motorized accessories are generally powered from high-amperage circuits, these narrow circuit board traces are susceptible to damage from excessive currents. Printed circuit traces function as wires carrying signals from one point to another. Depending on the cross-sectional area, they can carry only a certain amount of current before the heat generated by I^2R losses causes them to either melt or to become hot enough to delaminate, resulting in damage to the PCB and mounted components.

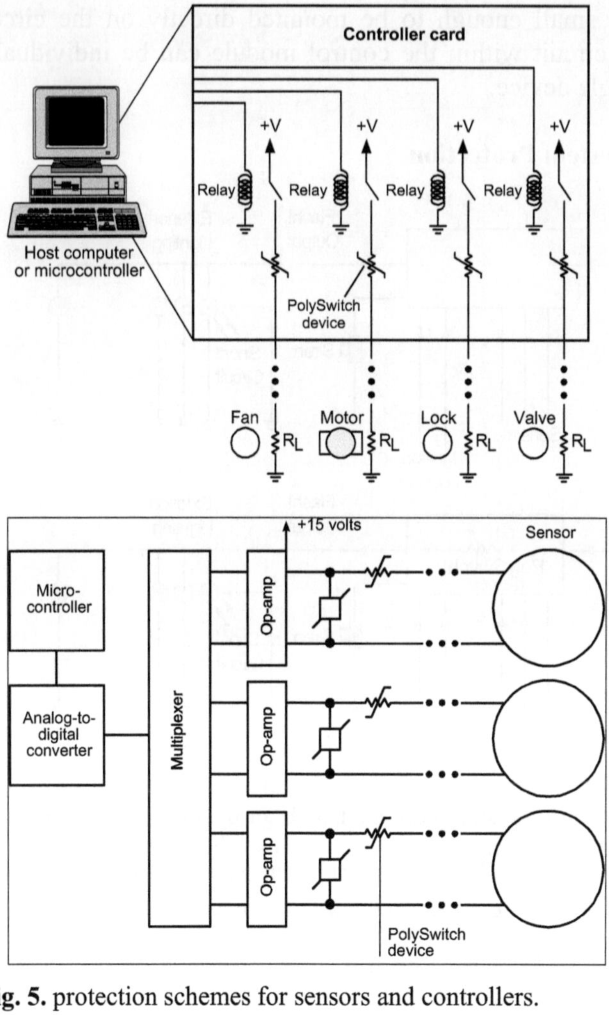

Fig. 5. protection schemes for sensors and controllers.

Electronic module outputs typically require protection from overcurrent situations caused by a short circuit, or high stall current of motors. They can also be damaged by failure of some other portion of the system, such as a diode short or loss of a power ground. Fuses are not considered an acceptable solution to these potential problems because they are one-use devices and must be replaced in the event of a transient fault. Multi-component circuits used to sense and switch, or SmartFETs, are frequently used in this application, but they require careful design, consume valuable board space, and can be quite costly.

PPTC circuit protection devices are gaining acceptance as a practical, cost-effective solution to overcurrent and/or overtemperature protection of electronic modules. Because they rapidly and effectively limit current to safe levels and are small enough to be mounted directly on the circuit board, each power circuit within the control module can be individually protected with a single device.

Vehicle Security System Protection

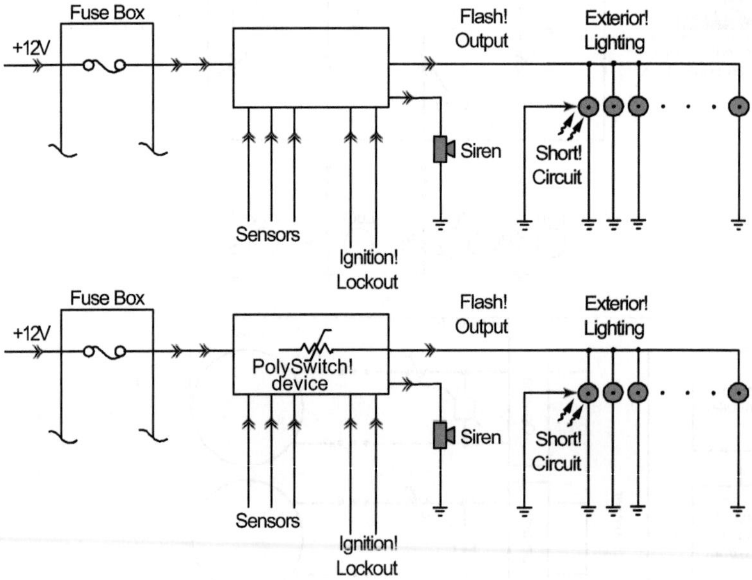

Fig. 6. In a vehicle security system equipped with an upstream fuse (top schematic), a short circuit in an exterior lamp blows the fuse, thereby disabling the entire security function. A PPTC device in series with the exterior lamp (bottom schematic), prevents the fuse from blowing and thus prevents the integrity of the security function.

Another application for polymeric PTC devices is in vehicle security systems. Such systems generally disable engine controls, sound an audible alary (the horn or a siren), and activate the exterior lighting flash (ELF) function to flash the marker lamps and headlights upon a security breach. Thieves can defect some security systems by simply removing the lens from an exterior light, then short-circuiting the lamp with aluminum foil. The short circuit blows the upstream fuse in the security system, disabling the system and allowing undetected entry into the vehicle.

As shown in Fig. 6., a PPTC device in series with the ELF circuit isolates any short circuit faults in the exterior lights. In selecting an appropriate PPTC device, it is necessary to coordinate the current requirements of the exterior lighting and the blow characteristics of the upstream fuse. Two important parameters in the selection process are the maximum ambient operating temperature and the nominal RMS current of the ELF function. It may be more appropriate to conduct empirical measurements on a typical system, than to use analytical techniques since multiplication of the duty cycle of the flash function and the nominal operating current of the lamp loads does not reflect the high inrush current of the incandescent lamps. Once the PPTC device is selected, it is necessary to select the upstream fuse such that the time-to-blow of the fuse is longer than the time-to-trip of the PPTC at operating temperatures.

Small Motor Protection

Most automotive actuators are used in applications that require them to move something until it reaches the end of its motion range – to move a seat or close a window, for example. However, since these activities can be manually controlled the actuator may remain energized after the mechanism reaches its limit of travel. When this occurs the actuator stalls and its back EMF (electromotive force) falls to zero. Without the back EMF opposing the supply voltage, the actuator's current may rise rapidly to levels typically between two and four times its normal operating value.

Because the actuator's winding is made with very small-gauge wire, the high stall current causes a rapid rise in temperature. Often within seconds the temperature may rise sufficiently to permanently damage the enamel varnish used to insulate the wire in the actuator's winding. With the loss of insulating properties, turn-to-turn short circuits may develop throughout the winding, rendering the actuator inoperable and creating the potential for a thermal event.

When the current and/or temperature of a winding rises above a certain value, the PPTC device latches into a high-resistance state, limiting current to a low level and preventing damage to the actuator. After the fault and power are removed and the PPTC device cools, the device resets to allow normal current flow.

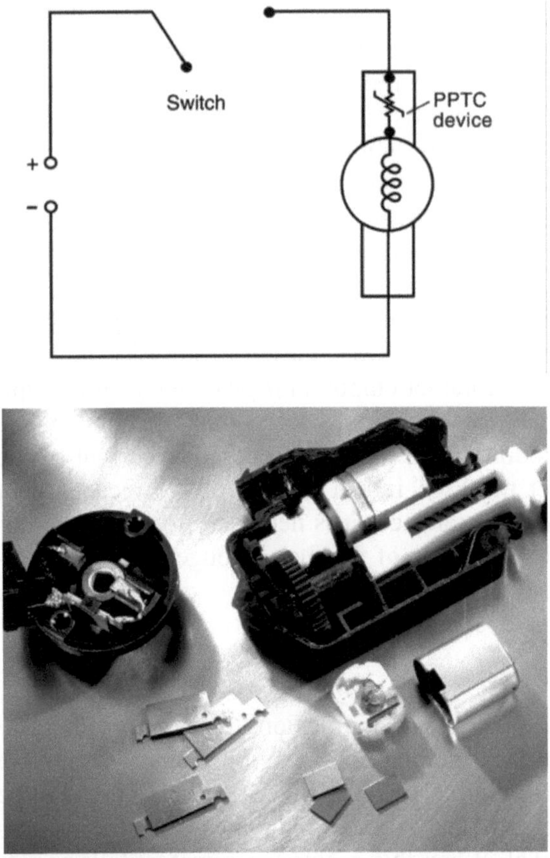

Fig. 7. To interrupt excessive current, PPTC devices are wired in series with the actuator windings.

5 Compliance with Industry Standards

Industry standards play an important role in the design of a vehicle's electrical/electronic system. The Automotive Electronics Council (www.aecouncil.com), a joint effort of Delphi Packard, Delco Electronic Systems, DaimlerChrysler and Visteon Automotive Systems, has published the AEC-Q200 Stress Test Qualification for Automotive Grade Passive Components. Most first and second tier automotive suppliers in North America have adopted this standard.

AEC-Q200 (Rev B) recently added test requirements for polymeric resettable circuit protection devices. The test plan includes seventeen electrical and environmental stress tests that require electrical verification tests prior to and after each stress. The electrical verification tests are designed to test that parts meet performance specifications for resistance, time-to-trip (TtT) and hold current at three different temperatures (-40 C, 25 C and max T).

Tyco Electronics, manufacturer of Raychem PolySwitch® PPTC devices, has developed test procedures that define performance limits prior to and after the qualification stress tests. The Raychem PS400 specification encompasses the AEC-Q200 standard. It incorporates relevant physical, functional, environmental, electrical, and mechanical requirements specified in a variety of ANSI, ISO, JEDEC, UL and military standards. A copy of this document is available upon request from the manufacturer.

Many of Raychem Circuit Protection's PolySwitch devices are qualified for and widely used in automotive designs. The PS400 test procedure was developed to provide compliance with the AEC-Q200 standard when it is required, and to simplify device specification – by assigning new model codes and markings that identify selected devices as PS400 compliant.

6 Summary

PPTC devices provide net cost savings through reduced component count and reduction in wire size. They can help provide protection against short circuits in wire traces and electronic components, in the current 14 V PowerNet, and the future 42V-PowerNet. The device's low resistance, relatively fast time-to-trip, and low profile improve reliability in a small footprint. Other benefits include manufacturing compatibility with high-volume electronics assembly techniques, and greater design flexibility through a wide range of product options.

New Fusing Concept under Consideration of Wire Characteristics

Werner Hinrichs, Jürgen Scheele, Ronald Zörn

PUDENZ GmbH, 27243 Dünsen, Germany

Abstract

Due to new requests of the automotive industry more and more electrical power must be transmited through the wiring harness. The power consumption of new automotive applications rises up rapidly. Solutions for these new board net architectures are the 42 V technology and new fusing concepts. At the present conventional protection systems are mainly specified on the consumer characteristics but not on the cable characteristics. So consequently the cables are often oversized and a higher current could flow through the cables.

The new fusing concept is focused on the protection of cables. To reach this requirement, it is necessary to move the time-current characteristics of the fuses very closed to the time-current characteristics of the used cables. This allows to use the cables up to their thermal limit without damaging by overloads or short circuits. With this fusing concept it is possible to decrease cable material and weight and to reduce the costs.

1 Introduction

In automotive applications the fuse protection of cables (wire harness) and electrical consumers constitutes a significant portion of the electrical system. At the present the cables and the consumer will be protected by special vehicle fuses with fixed rated currents. The current ratings are defined in specifications like DIN 72581 or ISO 8820. The decision for using a special fuse and cable in an application depends on the power consumption and maximum operating current of the connected consumer. For fuses the recommended rated current is ~125% of the operating current. The cable cross sections are also fixed and must be selected according to specifications under consideration of the ambient temperature and their fixed position. In a defined application consist of consumers, cables and fuses the cables often are able to transmit higher currents than the operating current because of their fixed cross sections.

2 Fuse protection in automotive applications

2.1 State of the art

In general fuse links are self-acting break appliances for the protection of electrical devices against unallowed current loads and short circuits. The current flow is interupted by the melting element of the fuse in which the current flows. The following Fig. 1. shows a variation of different fuses. They have different blade terminals, insulations (housings), colors and sizes. Each fuse type has favor application areas almost within the security systems of the vehicle electric. The choice for a fuse depends on the application, on the electrical load, on the continous current and on the ambient temperature.

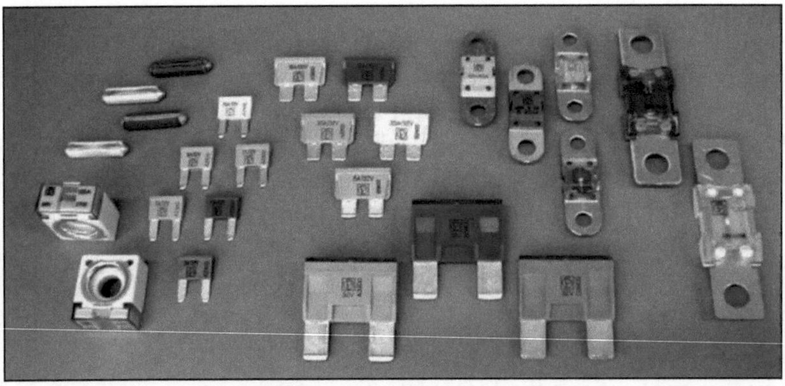

Fig. 1. Variation of PUDENZ fuses

For this miniature fuse links international regulations and recommendations (e.g. DIN 72581, ISO 8820, SAE) are valid. The rated current of a fuse link should approximately correspond to the operating current of the device or assembly unit which must be protected. The rated currents of fuse links are fixed and shown e.g. for the FK-fuses in Fig. 2.

Fuse-type:	Rated current in [A]															
	1	2	3	4	5	7,5	10	15	20	25	30	40	50	60	70	80
FK1	x	x	x	x	x	x	x	x								
FK2	x	x	x	x	x	x	x	x	x	x	x					
FK3									x		x	x	x	x	x	x

Fig. 2. Rated current values according to the DIN 72581

All types of fuse links can be selected by the same method. With regard to the product safety of the device and the life/reliability of the fuse links, a correct choice is important. Depending on the operating current of a consumer the necessary cable cross section must be selected, see Fig. 3., step 1. With the choosen cable cross section the corresponding fuse link can be selected (step 2). Additionally for the selection of a fuse link some general conditions just like electrical loads, transient currents and ambient temperatures must be taken into consideration (step 3).

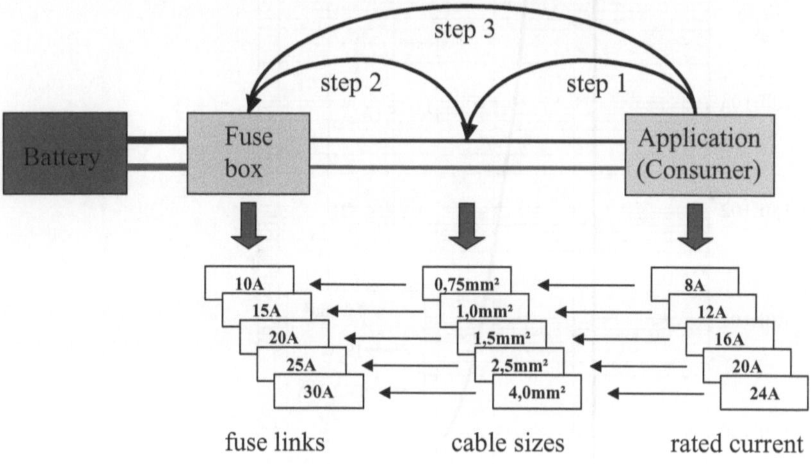

Fig. 3. Principle structure of a car electric system

Due to different specifications it is recommended that the operating current is equal or less than the rated current of a fuse link multiplied by the safety factor 0,8. The factor 0,8 is valid for e.g. the FK2 fuse links.

$$I_{operating} \leq I_{rated} * 0{,}8 \tag{1}$$

Is for example the operating current of an application max. 18 A and the cable max. longterm temperature 105°C, the necessary cable cross section must be 4 mm². The calculated rated current of the fuse link is 18 A/0,8 = 22,5 A. Now the selected fuse link must be a 25 A fuse link (e.g. a FK2 fuse link with a rated current of 25 A).

This example considers the arising cable temperatures, but the possible electrical loads and transient currents are disregarded. The time-current characteristic is determined by the min. and max. possible fault current and the max. allowable operating time. It becomes clear that the load limit of the 4 mm² cable isn`t reached. The cable is able to transmit a higher operating current than the 18 A. This situation is shown in Fig. 4.

The diagram of Fig. 4. shows the calculated and choosen time-current characteristic curves of the 4mm² cable, the FK2 25 A fuse link and the operating current 18 A.

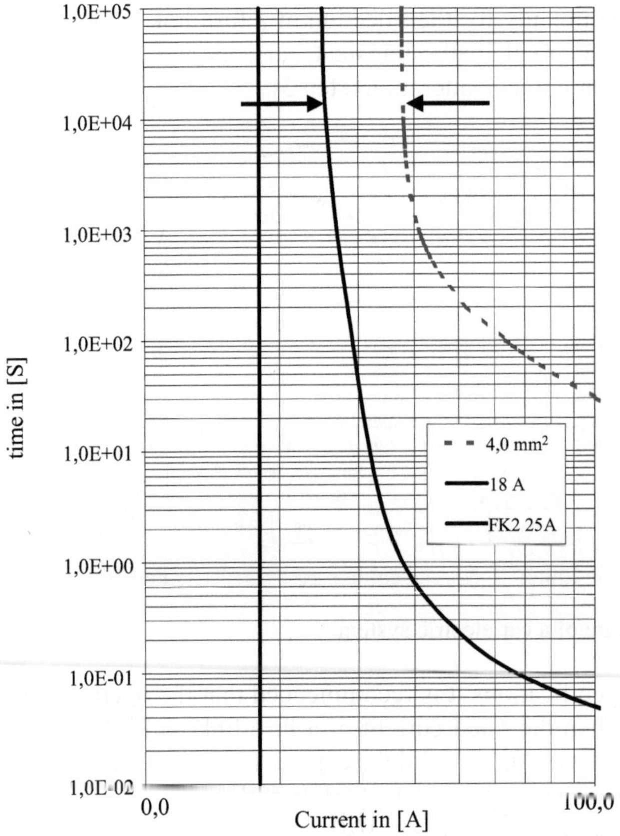

Fig. 4. Time-current characteristics of a FK2 25 A and a 4 mm² cable

The distance between the 4 mm² time-current characteristic of the cable and the time-current characteristic of the FK2 25 A fuse link is clear visible, see the arrows in Fig. 4. This unsatisfactory situation can be changed by the new fusing concept.

2.2 The new fusing concept

Based on the experience of conventional fusing systems new protection concepts especially for car applications arises, but they are still under development. The main task of the new fusing concept is comparable to the task of the conventional fuse links, it is to protect cables and consumer against overloads and short circuits.

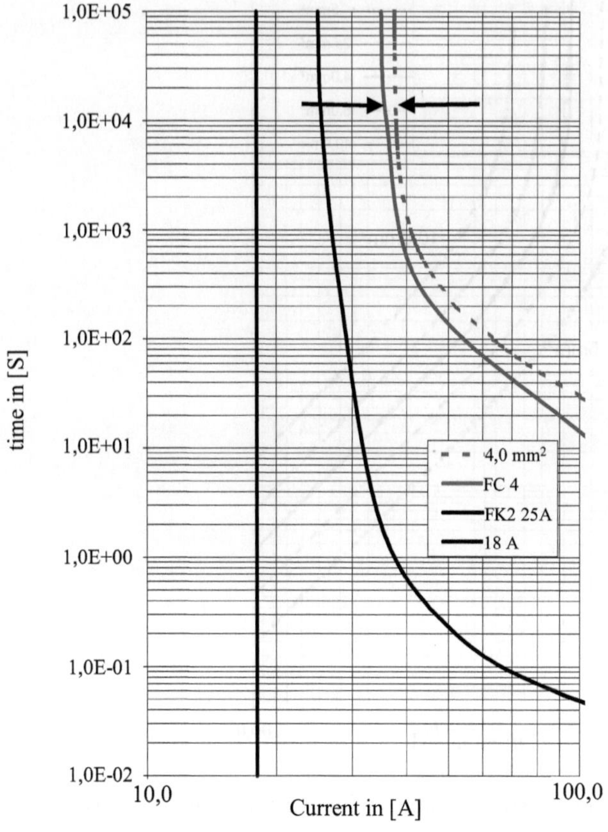

Fig. 5. Time-current characteristics of FK2, FC 4 and 4 mm² cable

The idea of the new fusing concept is to move the time-current characteristics of fuse links very closed to the time-current characteristics of the cables (Fig. 5.). That means the cables/cable insulations can be loaded up to their temperature limit. In this case a higher current can flow through the cable up to the temperature limit of the cable insulation (see arrows in-Fig. 5.), or on the other hand in an existing application the cable cross section can be choosen smaller, that results in less cable material/weight, lower costs and a more simple assembly. To make this concept reliable, the melting element must have a very exact time-current characteristic with small tolerances and a low cold resistance dispersion.

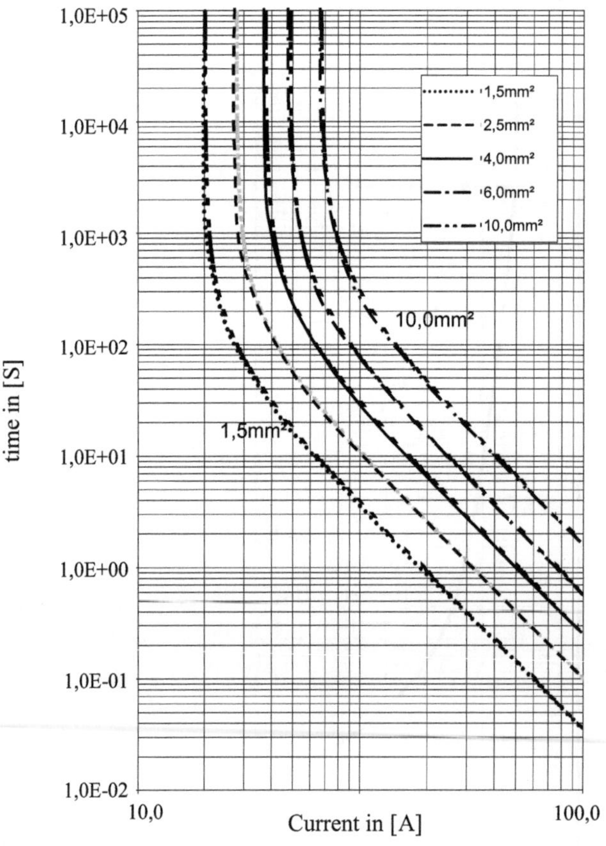

Fig. 6. Time-current characteristics of cables with different cross sections

For the development and analysis of cable protection the time-current characteristics must be prepared for each used cable. This can be performed by two different methods. One method is the practical measurement of time and current depending on the ambient temperature and the cable temperature limits. This is a very time consuming and costy method to get the typical curves.

The more efficient method is the calculation of the time-current curves by using a formula (see Form. 2). For the approximate calculation of the cables time-current characteristics (Fig. 6.) the specific resistances of the conductor materials and the types of insulation are important. Assumptions for the calculations are the ambient temperature (e.g. 65°C), the long-term temperature (e.g. 105°C), the short-term temperature (e.g. 145°C) and the delta temperatures (e.g. 40 K, 80 K).

A problem for the constant and reliability of the cables are gradual thermal destructions because of long duration overloads.

The resulting formula for the calculation of the different time-current characteristics is represented by the following formula. With this formula the currents will be calculated depending on the max. limited insulation temperature, the ambient temperature, the cable cross section, the cable time constant and the choosen time.

$$I(t) = \sqrt{\left[\frac{T_{insul\,\max} - T_A}{A * (1 - e^{\frac{-t}{\tau}})}\right]} \tag{2}$$

To guarantee cable protection it is necessary that the time-current characteristics of the fuse links do not intersect the typical time-current characteristics of the cables over the entire ambient temperature range specified. To simplify this view it is assumed that the ambient temperature for the cables is the same as for the fuse links. If the cable and fuse temperatures differ markedly, the cable characteristic must be compared with the typical characteristic of the fuse link at the lowest specified temperature. In this case the time-current characteristics of cable and fuse link are farther apart, because of their temperature dependence.

For this new fusing concept also all types of fuse links can be selected by the same method. It is in comparision to the selection method of the conventional fuse links more comfortable. Depending on the operating currents of the connected consumer the necessary cable cross section must be selected, see Fig. 3., step 1. With the choosen cable cross section the corresponding fuse link is automatecally defined Fig. 3., step 2). Is for example the choosen cable cross section 4mm², the type of fuse link is a FC 4. With this two steps the application is specified.

For the user it will not be necessary to consider the general conditions just like electrical loads, transient currents and ambient temperatures, because they are considered in the construction and development of the fuse link.

This new fuse links are called „FC“ fuse links and FC means „Fuse for Cable“.

2.3 Benchmark

To get a clear overview of the differences between conventional fuse links and the new FC fuse links a comparision is necessary. In order to show general property strenghtenes and weaknesses the fuse links are listed in a valuation table by means of some relevant criteria (see Fig. 7.). For an objective comparision between the FC fuse link and conventional fuse links representative the FK3 fuse link and the FJ3 fuse link are selected.

	Application areas	Structure	Fitting area	Installed height	Utilization factor	Costs	Future
FK3 fuse	o	o	-	o	-	+	o
FJ3 fuse	o	o	o	o	-	o	o
FC fuse	+	+	+	-	+	?	+

Fig. 7. Valuation between the fuse links FK3, FJ3 and FC

The FC fuse link is developed for applications with max. 58 V rated voltage and that's why the FC fuse link is also optimal for the 42 V technology. The structure of the FC fuse link is simple. It consists of 3 single parts, the cover, the insulator and the one piece metal part. The metal part consist of the blade terminals and the melting element.

The used fitting area of the FC fuse link is in consideration to other fuse links (rated current/voltage) very small and the installed height of the FC fuse link is comparable to the installed height of standard automotive relay`s. The utilization factor of the FC fuse link is in the opposite of conventional fuse links very high, because the FC characteristics and properties based on characteristics and properties of the corresponding cables. The prices between the different fuse link types differ immense.

This new FC fusing concept reaches interest by the automotive industry and it is absolutely conceivable that this kind of fuses will be used beside conventional fuse links in future board net architectures.

2.4 Areas of applications

In a first step it is planned to develop the new FC fuse links according to the table of Fig. 8. In a second step the development of further FC fuse link types will follow. The max. rated voltage of each FC fuse is 58 V.

FC fuse application				
Fuse name	Cable cross section	Insulation color code	Cover color code	Nominal voltage
FC 1,5	1,5mm²	blue	grey	58V
FC 2,5	2,5mm²	yellow	grey	58V
FC 4	4,0mm²	green	grey	58V
FC 6	6,0mm²	black	grey	58V
FC 10	10,0mm²	red	grey	58V

Fig. 8. Applications for the FC fuses

2.5 The FC fuse – state of development

The development of the FC fuse link is focused on the optimized protection of cables under consideration of their time-current characteristics. The FC fuse links will be developed for different cable cross sections (Fig. 8.). The FC shape, the marking and the one piece metal part of the FC fuse link is shown in Fig. 9.

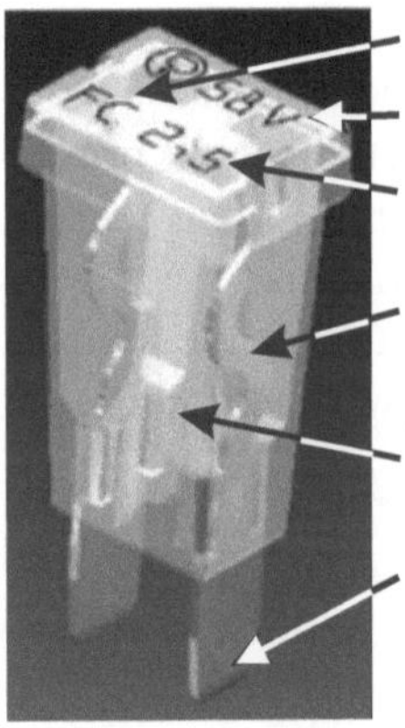

Fig. 9. FC fuse link with the significant features

2.6 Summary

New requests of future automotive applications are responsible for the development of new protection systems. A lot of conventional protection systems aren`t able to meet the requirements of e.g. the 42 V technology, high current transmissions and high power consumtions. One future fusing concept is the fusing concept which considers the cable time-current characteristics. It is realized by moving the time-current characteristic of the fuse link very closed to the time-current characteristic of the cable. The cables and mainly the cables insulation now can be loaded up to their thermal limits without any damages. On the other hand it is possible to reduce the cable cross section depending on the specific application. This possibility reduces the material amount, the weight and for sure the costs. The name of this new fuse link is called „FC".

3 Definitions, Acronyms, Abbreviations

FK-fuses:	Flachkörper
FC:	Fuse for Cable

4 References

[1] DIN 72581-3, Sicherungen für Kleinspannungsanlagen, Beuth Verlag GmbH, Berlin, März 2001
[2] ISO 8820-3 (draft version), Road vehicles – Fuses – Part 3:, Fuse link with tabs (blade type), ISO central Secretariat, Geneva, September 2000
[3] Weber M, Molex Elektronik GmbH, Kabelbäume im Automobil, Auto&Elektronik, A & E, März 2000
[4] Wilhelm PUDENZ GmbH, Internal knowledge and experience, research & development, reports and documentations

Overvoltage Protection Devices for the Automotive Power Network

Paddy O´Shea

Vishay Intertechnology Inc., Ireland

Erich Niebler

Vishay Electronic GmbH, Selb

1 Introduction

Many years ago the automobile began without electronic components. It was a simple vehicle with a minimum of electrical components and features for creature comfort. Today's automobile is a vastly different generation of vehicle, which includes electronic modules for control, fuel efficiency, safety, entertainment and comfort. We live in a world, which demands ever-increasing features, quality, and reliability from our automobiles. These customer demands have increased the power requirement in vehicles to a level where the traditional 14 V/28 V systems are struggling to keep up. Moving to a 42V-PowerNet allows the automotive industry manufacture vehicles, which can power the increasing electrical loads.

To enable vehicles meet the reliability and safety requirements Overvoltage Protection Standards are in place, such as

- ISO 7637, electrical disturbance by conduction and coupling,
- DIN40839, electromagnetic compatibility (EMC) in road vehicles,
- WGS/WD 03/2000, conditions for electrical and electronic equipment for a 42V-PowerNet and
- ISO/WD21848, electrical and electronic equipment for a 42V-PowerNet

Manufacturers of automobile systems must produce systems, which meet the Overvoltage Protection Standards reliably requirements over a long period of time. To maximise the supplier's capability to meet these reliability and legal requirements the attitude being taken by the industry is "The Originator of transient voltages must prevent and protect". Some modules generate transient voltages and the manufacturer is therefore responsible to ensure the excess voltages are limited below specified maximum limits and the excess energy safely dissipated within the module.

The 42V-PowerNet will requires the use of Transient Voltage Suppression (TVS) products with increased clamping capabilities. This is because the alternator power output will continue to rise as the level of electronic and electrical modules in automobiles increases. Also the voltage gap between the normal operating voltage and the maximum specified voltage is much smaller than that with the 14 V systems. See voltage clamping, sections, top, of Fig. 1.

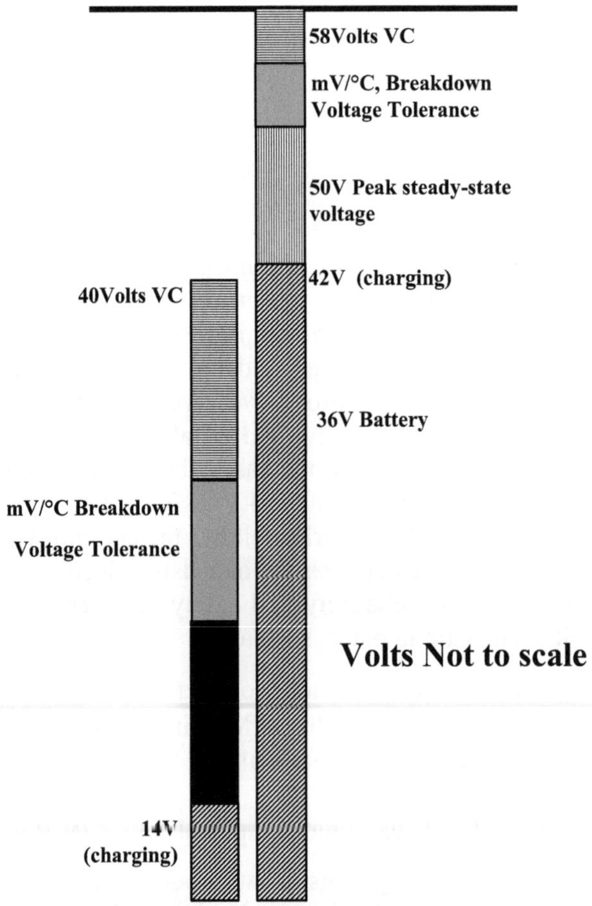

Fig. 1. 14 V/42 Vdifferences & Challenges

2 What is the threat on 14 V/28 V/42 V systems

The threat to electronic systems comes from high and low energy transient voltages. High-energy transient voltages come from load dump, field decay with load changes and 42 V to 12 V short circuits. Low-energy transient voltages are caused by ESD and relay / electronic load switching.

The most severe automotive transient voltage is the Load Dump. When the alternator load, battery, is disconnected at high charging current the high impedance forces the alternator voltage to increase. This excessive voltage creates a discharge path and dissipates the energy through the electronic circuit. This pulse is typically **100 ms – 500 ms** duration. TVS-devices, which limit the peak voltages, are designed to absorb this high-energy pulse safely by diverting the transient energy away from the more sensitive devices.

3 Load Dump Solutions

The ideal attributes of a Transient Voltage Suppressor (TVS) component are

• Dissipate high energy pulses reliably	**PPM**
• Wide operating temperature range	**-65°C to +185°C**
• Very low thermal impedance	**W/°C**
• Very low voltage temperature variation	**mV/°C (%/°C)**
• Very good clamping factor	**V_c/V_{BR}**

Fig. 2. SM8S Series

The increased performance requirement and the tighter specification of the 42V-PowerNet will stretch the above attributes to the limit because the operating window is considerably smaller than with the 14 V power system. Device impedance, temperature variation (coefficient), operating temperature range, and the 58 V clamping limit present a serious challenge to effective transient voltage suppression technologies.

Device impedance causes the voltage to increase as current through the device increases. As the ambient engine temperature increases the devices power handling capability decreases. Voltage temperature variation causes the breakdown voltage of devices to increase or fall with temperature changes. Manufacturing and voltage grading tolerances also affect the voltage range available for use in this 42 V application.

After all of the above factors, the TVS device breakdown voltage must be greater than the battery/alternator voltage at minimum ambient temperature and have a low enough impedance to limit the voltage below 58 V at peak current while dissipating as much power as possible at a high ambient temperature.

There are a number of protection topographies available to give protection against load dump within the very tight limits of the 42V-PowerNet specification as shown in Fig. 2. and Fig. 3.

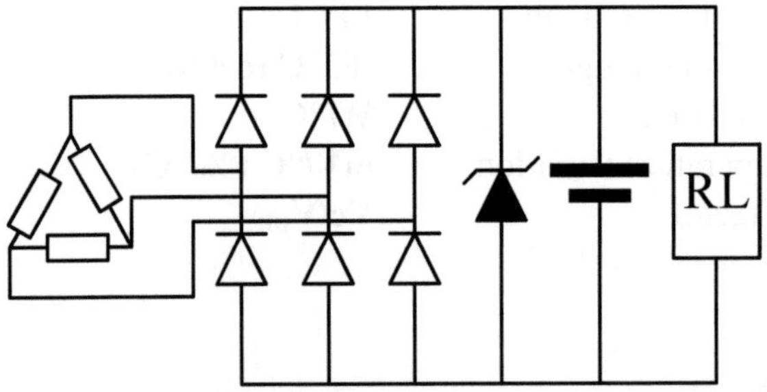

Fig. 3. Shunt Configuration

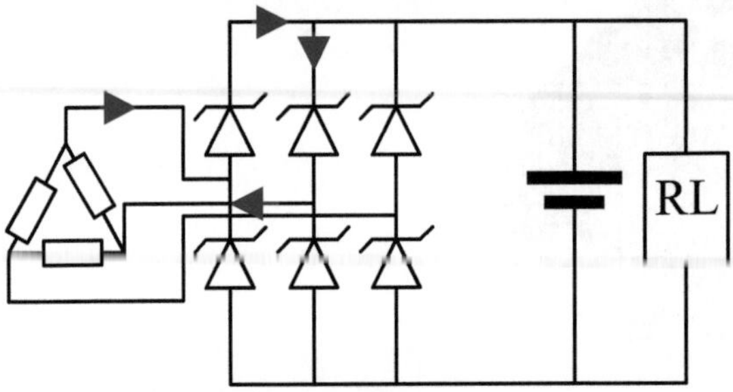

Fig. 4. Integrated Avalanche Rectifier

Fig. 3. shows the traditional shunt configuration for load dump protection, which can also be applied to the 42V-PowerNet. The TVS device is placed directly across the alternator output. Tight voltage grading is required for the device in 42 V systems. Very good thermal impedance characteristics and a silicon area, which can handle the high current for the 100 ms – 500 ms duration, are required also. To improve power-handling capability at the high ambient temperatures and reduce clamping voltage, devices can be stacked in series.

Fig. 4. shows the bridge rectifier using Avalanche-rectifiers. This configuration spreads the load dump across the six diodes. This is done to increase the voltage selection range, ease the device impedance limitations and reduce the transient power dissipated in each diode. Because the breakdown voltage is lower than the peak output voltage of the alternator, high reverse current will flow for a time causing the devices junction to heat. This increases the breakdown voltage to a level higher than the peak voltage and it is then self-regulating. This configuration also minimises the component count.

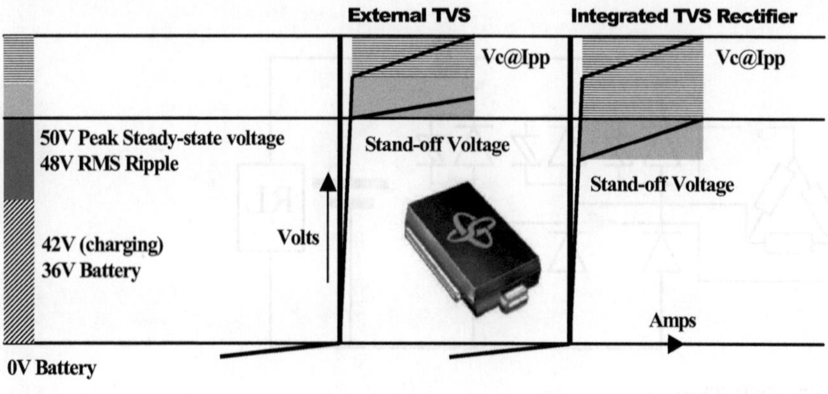

Fig. 5. I/V curves for shunt and integrated configurations

Fig. 5. shows the I/V curve for both configurations. It also highlights the very limited voltage ranges available for device selection and voltage clamping at high current over the wide operating temperature range. The voltage range within which the protection devices must operate is very different from the 14 V system and poses a real challenge to the system designers.

Fig. 6. shows a FET TVS device where the diode chain sets the clamping voltage and the transient current flows through the FET. The diode breakdown voltage is diffused so that the positive and negative temperature coefficients of the reverse and forward biased diodes are equal and cancel (0 mV/°C). Great care needs to be used when selecting this option as not having a positive temperature coefficient prevents other TVS components in the electrical system from being called in to share the transient load.

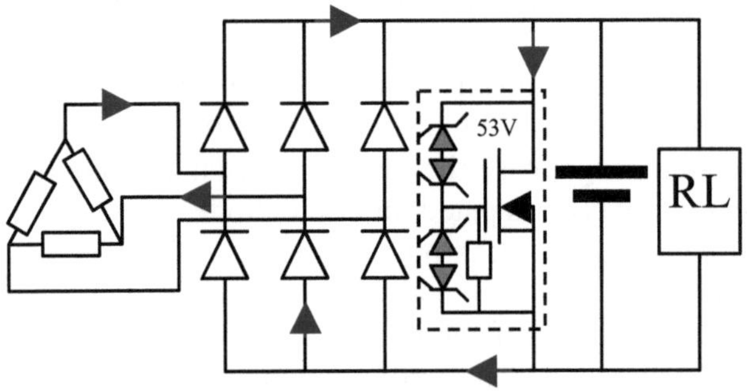

Fig. 6. FET-TVS

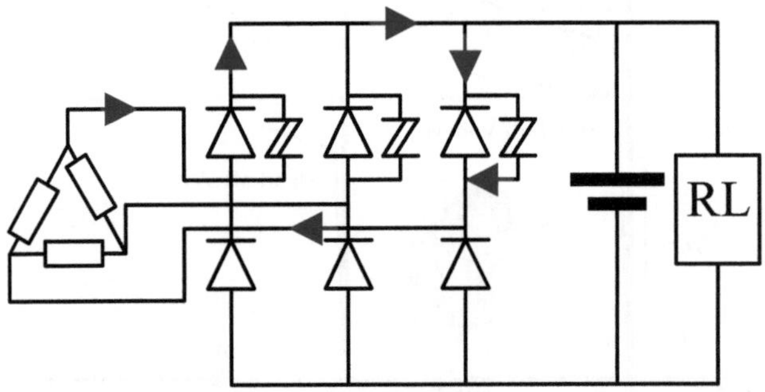

Fig. 7. TVS-Thrystor

Fig. 7. shows TVS-Thyristors connected into the bridge rectifier. When the alter-nator voltage increases above the Breakover Voltage, V_{BO}, current flows in the TVS-Thyristor switching it into its low impedance, low voltage state. The alternator windings are shorted and the transient energy is dissipated across the windings as heat rather than in the silicon. The fast switching time, 1 µs – 2 µs, prevents the silicon from heating and increasing the peak voltage so the clamping factor is very low.

Fig. 8. shows the I V curves for the TVS-FET and TVS-Thyristors.

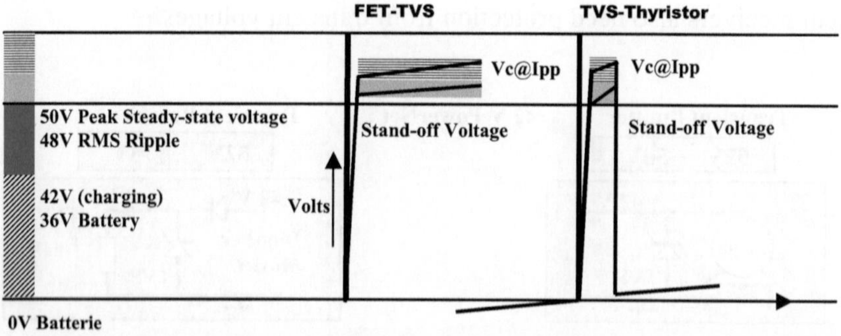

Fig. 8. I/V curves for TVS-FET & TVS-Thyristor Configurations

4 Module Level Protection 12 V/14 V/42 V

Module level protection protects electronic systems distributed throughout the vehicle against the following threats

- Load dump
- Failed regulator
- Alternator field decay transient
- Reverse polarity connection
- Inductive load switching
- Electrostatic discharge

Effective protection against transient voltages requires a combination of good grounding, bonding, shielding, protective devices, and over-current limiting devices e.g. fuse. Series and shunt protection configurations are available with shunt being the most common. In normal operation series protection has low impedance, which increases as the input voltage increases. Shunt protection limits excess voltages and diverts transient current away from the circuit. Electrical or electronic modules are either transient emitters or receivers. See Fig. 9. & Fig. 10.

Any electrical or electronic module, which generates transient voltages, must include transient voltage suppression to limit transient voltages which may be emitted from the unit. This means by definition that there should be no transient voltages above 58 V transmitted into the power network. These transmitter modules are also vulnerable to load dump and other external transient voltages. However there could be high transient voltages generated from other sources, like intermittent contacts. Therefore transient receivers also need protection from transient voltages.

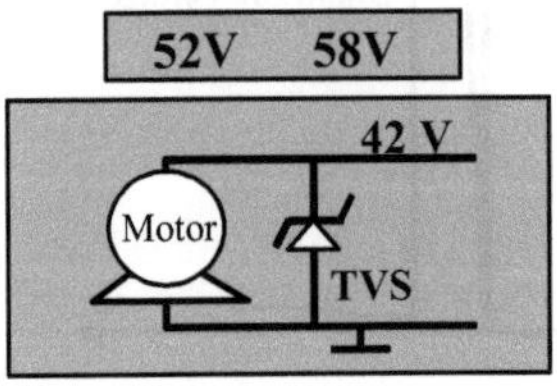

42 V PowerNet :

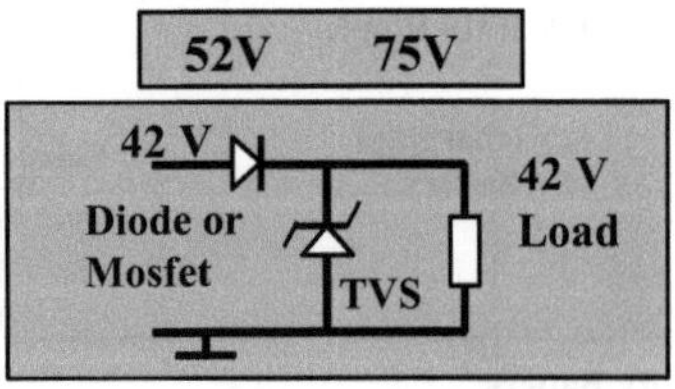

Fig. 9. Transient Voltage Emitter

Fig. 10. Transient Voltage Receiver

While transient receivers do not generate transient voltages they are vulnerable to transients from external sources. PTC heaters in new diesel engines are a good example of transient receivers. They are pure resistive loads that do not generate inductive transient voltages. However they are switched by MOSFETS. If the drain-source voltage,V_{DS}, is exceeded by a voltage from an external source they will fail. Many FETS used in these applications are designed with 75 V technology, so protection is required. Telematic systems and car radios are also examples of transient receivers. Even more sensitive to transient voltages are micro-controllers and memory components with 3 V to 5 V operating voltages.

4.1 Overvoltage & Reverse Polarity Protection

Where high power is not required 14 V modules and components will continue to be widely used in the automotive industry. The 14 V supply will be generated from the 42 V system via DC/DC converters. Within the 42V-PowerNet system one of the severest risks comes from shorting the 42 V power system to the 14 V system. See Fig. 11. 14 V modules are currently designed to withstand 24 V for one minute. High power TVS devices must limit the voltage to the 14 V power systems and divert the high current until the fusing on the 42 V power system isolates the short circuit load.

Another major risk within 42 V/14 V systems is the loss of the common ground connection. When this happens the ground point will float and the 42 V system will attempt to discharge into the 14 V power system. The 14 V TVS diode will be forward biased and protect the 14 V circuits from the reverse voltage. 14 V systems are specified for –2 V maximumreverse voltage. The series diode in the 14 V module will block the current preventing the 42 V system from discharging into the 14 V power system. See Fig. 12.

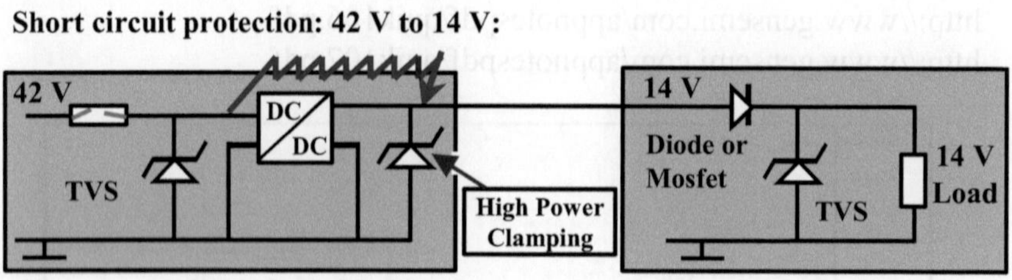

Fig. 11. 42 V to 14 V short cct. protection

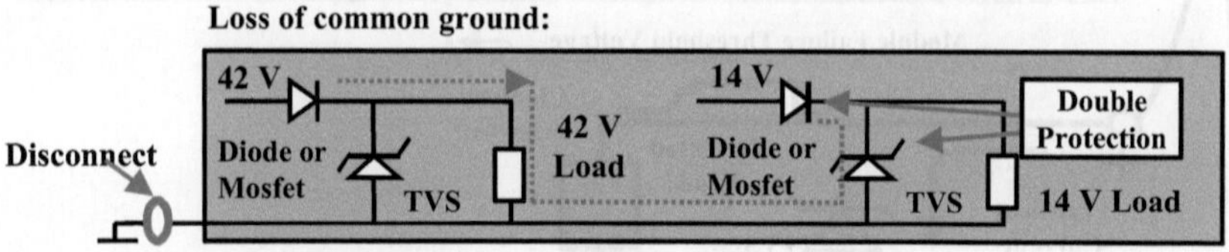

Fig. 12. Loss of common ground

4.2 Device Selection

Selection of a TVS device for an application comes from matching the device parameters with the operating voltage, operating temperature range, the peak pulse power and duration.

Fig. 13. illustrates these parameters in relation to the I/V curve of a TVS-Diode. Care needs to be taken when designing components into applications with high duty cycles and elevated temperatures. They may need to be derated.

For further detailed information email: Erich.Niebler@vishay.com

The power handling capability of a device can be increased and the impedance reduced by using series and parallel combinations of TVS diodes. These combinations maximise the power handling capability while giving the best clamping factor. The peak surge current through a series stack of TVS diodes is equal to the lowest specified peak surge current, I_P, of the TVS chain. The breakdown voltage is the sum of the individual breakdown voltages of the devices in the chain. When TVS diodes are connected in parallel the breakdown voltage needs to be graded to a tight tolerance to ensure all the devices share the surge current. If the devices are not closely matched the surge current will flow through the lowest voltage device and cause it to fail. For more detailed information on component matching please see:

http://www.gensemi.com/appnotespdf/quik106.pdf
http://www.gensemi.com/appnotespdf/quik107.pdf

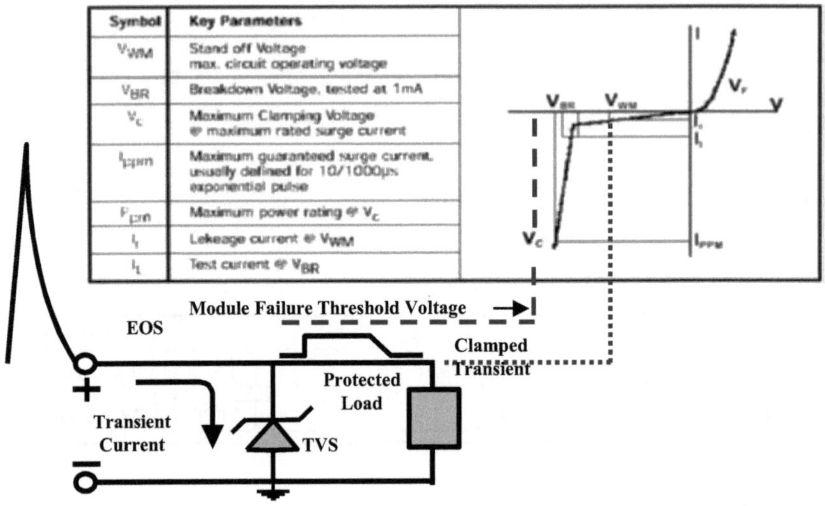

Fig. 13. Device selection and device parameters

5 Module Level Protection 12 V/14 V/42 V

Many people think ESD is a threat during the assembly stages of an automobile or its component parts. However the reality is that ESD is a threat during the lifetime of the vehicle. Vulnerable points are control panels, indicators, and accessible I/O pins. When electronic components are exposed to ESD they may not fail immediately or show any immediate signs of degradation. However they will continue to degrade to a point of total failure.

As semiconductor technologies advance with reduced geometries the electronic components are more sensitive to ESD. To provide effective ESD protection ESD devices need to have

- Very fast response times
- Low clamping and operating voltages
- Handle high peak ESD currents

- Survive repeated ESD strikes
- Minimal package size
- Minimal reverse leakage current

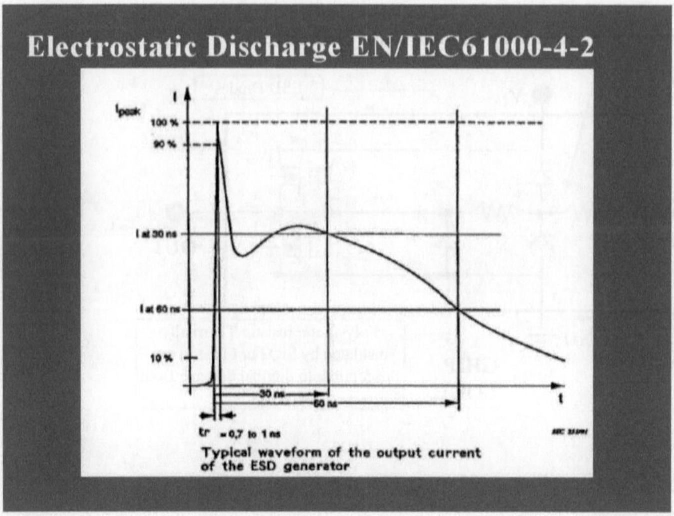

Fig. 14. Electrostatic discharge

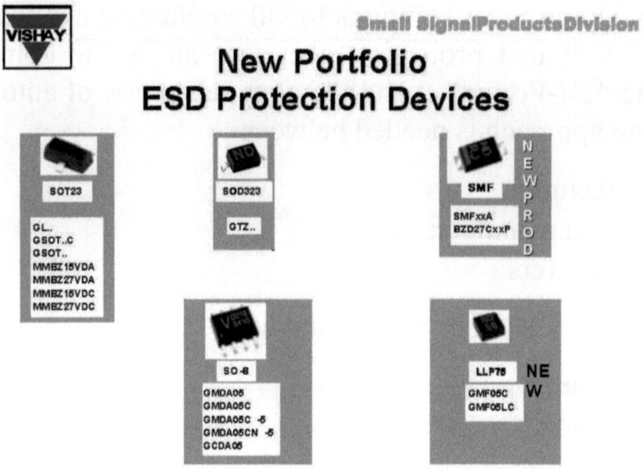

Peak pulse power of 200 **mW - 750 mW**
30 kV air discharge

Unidirectional & bidirectional options Bus and ESD input protection

Fig. 15. Portfolio ESD protection devices

While many ICs contain ESD protection, the level of protection must be checked against the level of exposure and the ESD specifications, which apply. Where appropriate additional protection must be provided. See Fig. 16. A wide variety of ESD products are available.

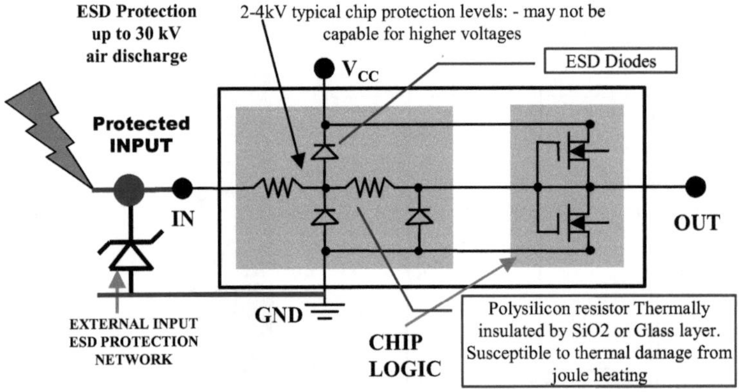

Fig. 16. External ESD protection

6 Summary

Vishay Intertechnology provides solutions for all automotive applications.

To provide an intergrated protection plan that meets the demanding specifications of the 42V-PowerNet and the next generation of automobile a system engineering approach is needed between

- Automobile manufacturers
- Module designers and manufacturers,
- Component manufacturers

The best overall system level protection includes

- Load dump protection
- Module level protection
- ESD protection

Transient Voltage Suppression devices need to be optimised to meet the increased performance requirements and tighter specifications in the 42V-PowerNet.

Power Switches for the 42V-PowerNet Solution for Power Net Protection and Applications

Matthias Kroeker

Tyco Electronics AMP GmbH, Paulsternstrasse 26, 13629 Berlin

Abstract

The demand for more electrical power in future cars is the motivation to think about a change of the voltage level from 14 VDC to 42 VDC. The more of available power, together with the possibility of steady arcing the potential risks increase. The dissipation of power within an arc could cause serious damage, e.g. the ignition of flammable material. Therefore not only the loads have to be adapted to the new power level, but also does one have to think about new safety and protection strategies. We would like to present our approach of arc-detection and leakage current monitoring. But to detect the fault condition is only the first half of the solution. If an error occurs, there has to be some device capable of separating the fault path and only the fault path from the rest of the power net. In this paper we will show how we had made our relays capable of this task. A possible power net protecting system could be build with these relays together with a combination of an arcing sensor and a leakage current detector. If the detection time is sufficiently short, the breaking capacity of these relays is remarkable. With the `tyco` µK relay we successfully switched several times a 520 A short circuit peak. This was achieved even though this relay is not bigger than a lump of sugar and has a continuos current rating of 30 A.

1 Introduction

Concerning the problem to switch off inductive loads Prof. Rieder wrote the following in his lecture notes. “An optimal DC switch has to consist of a resistor with an adequate capacity, by which the resistance...rises from zero to infinity and which...is regulated in a way, that the allowed overvoltage...will be just fully utilized but not exceeded.

The best, economical, safe and self regulating „„ on anytime available switching resistor with all the required properties is the burning arc, which when the contact opens ...spontaneously arise, limiting the switching over voltage,...and finally completely looses it's conductivity " [1]

In other words, when an inductive load path is disconnected from the power supply, the inductance will be de-energized by the occurring arc. For currents below 50 A up 80 A the arc resistance has a negative differential component, i.e. the voltage current product across the arc will adjust itself in a way that the input power just compensates the loss due to radiation and heat transport. For higher currents the arc resistance has a positive characteristic.

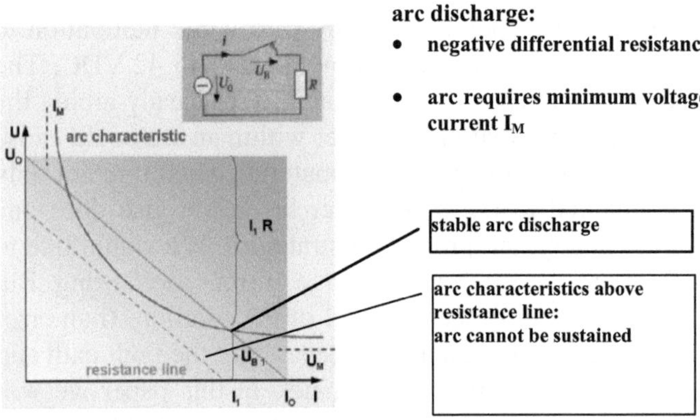

Fig. 1. Low Current Schematic VI diagram

These ideal qualities apply for all switching applications below 24 V so are turned at 42 VDC into the opposite.

The voltage drop across an opening contact is mainly determined by the anode and cathode fall. This minimum voltage U_m emerge at the moment the contact opens. For the free in air burning arc at normal pressure, U_m depends only on the electrode materials. At normal conditions the value for U_m is between 12 VDC for silver and 20 VDC for carbon[2][3]. High loads and supply voltages above U_m are the adequate and necessary conditions for arcing. The temperature of the anode and cathode region are only limited by the boiling temperature of the contact material, i.e. over 2.000 K. The temperature of the arc column can rise above 12.000 K [4][5]. At it's best, the electrode material evaporates and as a consequence the increasing arc length lets the plasma channel break down before the arc can set inflammable material on fire. Nevertheless the area where the arc occurs will be destroyed.

Considering this, it is necessary take measures against arcing. In contrast to the 14 V environment

1. tThe contact arrangements of switches and relays have to be modified to withstand higher voltages.
2. Precautionary actions have to be taken, that intentional or incidental disconnection of a power line does not lead neither to fire nor to a complete destruction of the contact area.

2 Switching 42 Volts Loads

The maximum voltage, which can be switched by a single contact pair depends on the characteristic arc voltage. As sketched below the arc voltage is a combination of the material dependent anode and cathode fall and the voltage drop across the column. The arc voltage has a decreasing characteristic up to 50 A and rises again for higher currents.

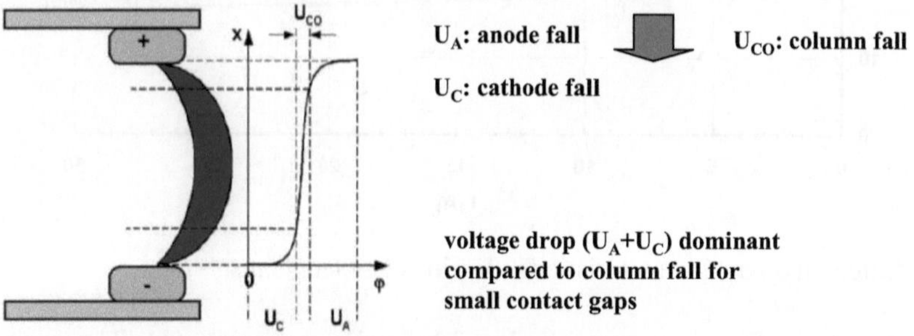

Fig. 2. Contributions to the Arc Voltage Drop

2.1 Influence of the Contact Gap

There are a lot of different empirical approaches concerning the functional dependencies between arc voltage and arc length. [6] [7] The arc length of a single contact in case of small contact gaps (s<1,5 mm) and low currents (I<5 A) could be described as

$$s = 3{,}85 * 10^{-3} * (U - Um)^{1.57} * I^{0.49} \quad [8] \qquad (1.)$$

for silver-cadmium contacts.

Although equation (1.) was only given for currents up to 5 A, it describes the arc characteristic satisfyingly up to 70 A.

The calculated arc limit curves for different contact gaps are shown in Fig. 3. The minimum contact gap for a given current rating can be constructed by drawing a straight line, which represents the resistive load. This resistance line starts at supply voltage and 0 A and ends for the voltage drop 0 V at full current load. The minimum contact gap is represented by the arc limiting curve, which just touches the resistance line but has no intersection with it. [9]

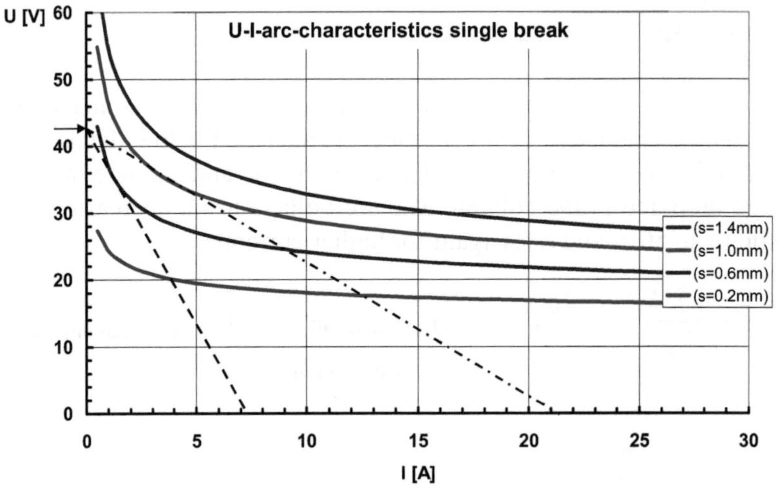

Fig. 3. Calculated Arc Characteristic for Various Contact Gaps

The traces in Fig. 3. were calculated by using equation (1). The resistance lines for a 7 A and a 21 A load were added. The switching capability limits derived from this diagram are in good accordance with results from experiments [7].

2.2 Double Break Contacts

As shown, the dependents between the arc voltage and the arc length is weak. It is difficult to enable a standard relay capable of 14 VDC to switch 42 VDC, only by increasing the contact gap. Relays in automotive applications normally have contact gaps between 0.2 mm and 0.6 mm for currents up to 70 A. As Fig. 3. indicates the limit current for a 0.6 mm contact width will be 7 A. For 50 A the gap has to be 1.4 mm. To double the working gap means to double effective magnetic energy of the motor coil. With coil currents kept constant a rough estimation shows, that the relay volume have to be enlarged by factor square root of 8. This is in contrary to the demand, that switching devices have to get smaller and smaller.

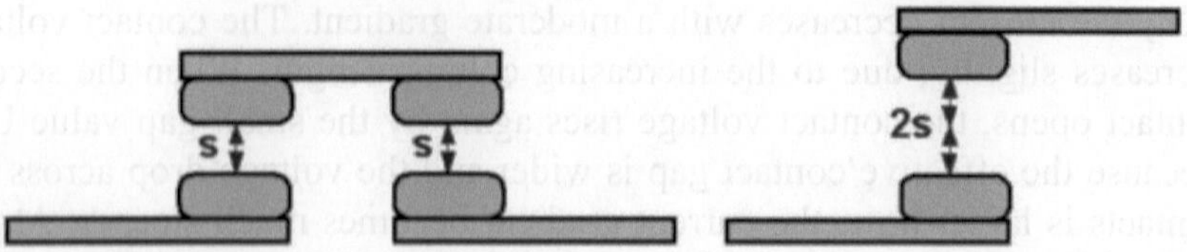

Fig. 4. Sketch of Single and Double Break Contact Arrangement

A double breaking contact consists of two simultaneously opening pairs of contacts. Obviously two contact gaps double the sum of the column length. Hence the arcing voltage for short contact gaps is mainly determined by anode and cathode fall, the advantage of a multiple breaking contact is much higher, than expected only for the increasing contact gap.

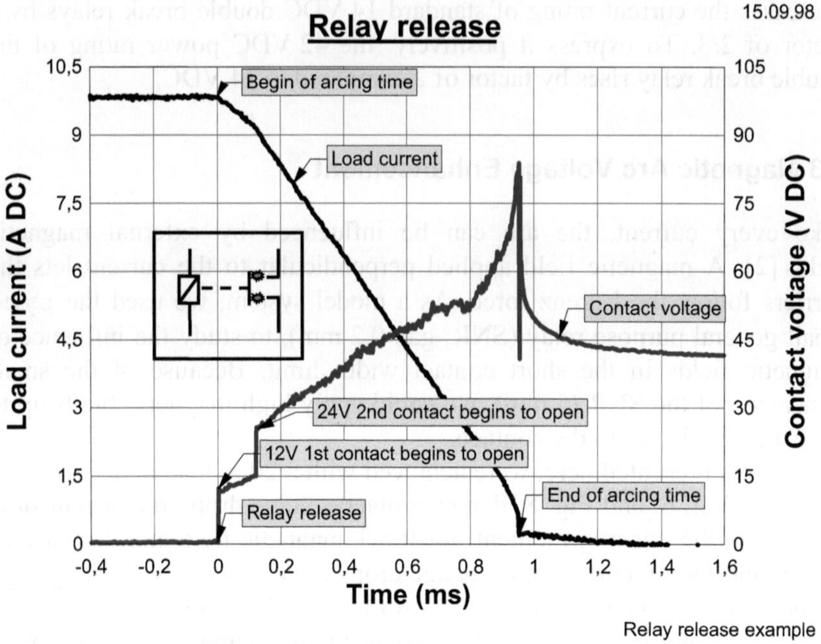

Fig. 5. Switching Behavior Double break contact

The diagram above shows the contact voltage and current flow, measured for an inductive load of 10 mH at the moment, when a double break (form U) relay opens. Because of mechanical imperfections, one contact opens with 60 μs offset. At first the contact voltage rises up to the characteristic small gap value U_m, i.e. 12 VDC for silver contacts. Simultaneously the current decreases with a moderate gradient. The contact voltage increases slightly, due to the increasing column length. When the second contact opens, the contact voltage rises again by the small gap value U_m. Because the effective contact gap is wider and the voltage drop across the contacts is larger now, the current gradient becomes much steeper. At 4.5 A the contact voltage and supply voltage are equal. At this point, the arc is exclusively feeded by the energy stored in the inductance. After 0.8 ms the contact movement comes to an end and the opening width has reached it's maximum. Then the contact voltage rises very fast and the arc extinguishes.

A supply voltage increased by the factor of 3, means that the power dissipation within the contact chamber has to rise a factor of 9. Therefore the contact erosion will be higher than at 14 VDC level. The current rating of these 14 VDC relays have to be adjusted to the new level. It can be taken as a rule of thumb, that the increase of the supply voltage by a factor of 3, decreases the current rating of standard 14 VDC double break relays by a factor of 2/3. To express it positively, the 42 VDC power rating of the double break relay rises by factor of 2 compared to 14 VDC.

2.3 Magnetic Arc Voltage Enhancement

Like every current, the arc can be influenced by external magnetic fields [2]. A magnetic field applied perpendicular to the current lets the carriers follow the Lorenz force. As a model system, we used the `tyco` small general purpose relay (SNR; gap 0.2 mm), to study the influence of magnetic fields in the short contact width limit. Because of the small dimension of the SNR (5 mm), we could apply high magnetic fields up to 0.7 T perpendicular to the contacts.

The data presented here, were achieved with a 20 A load and 10 mH inductance. Fig. 6. and Fig. 8. depict contact voltage drop and current flow vs. time in the opening moment, for 0 mT magnetic field and 700 mT. In Fig. 7. and Fig. 9. current and voltage drop are combined as V-I diagrams. In the case of 0 magnetic field the V-I diagram corresponds, as expected, with the arc limiting curve for small contact gaps. Below 60 mT, we found only a small influence of the magnetic field strength on the arcing behavior. Above 60 mT the arc limiting curve rises dramatically.

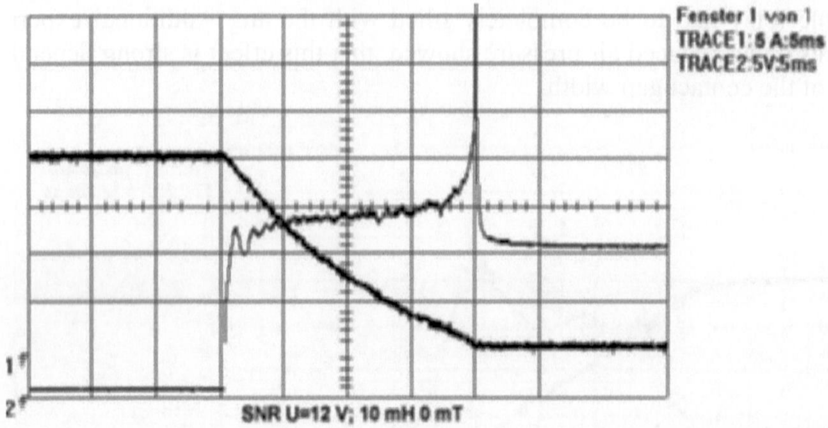

Fig. 6. SNR contact voltage and current inductive load 20 A, 10 mH B=O

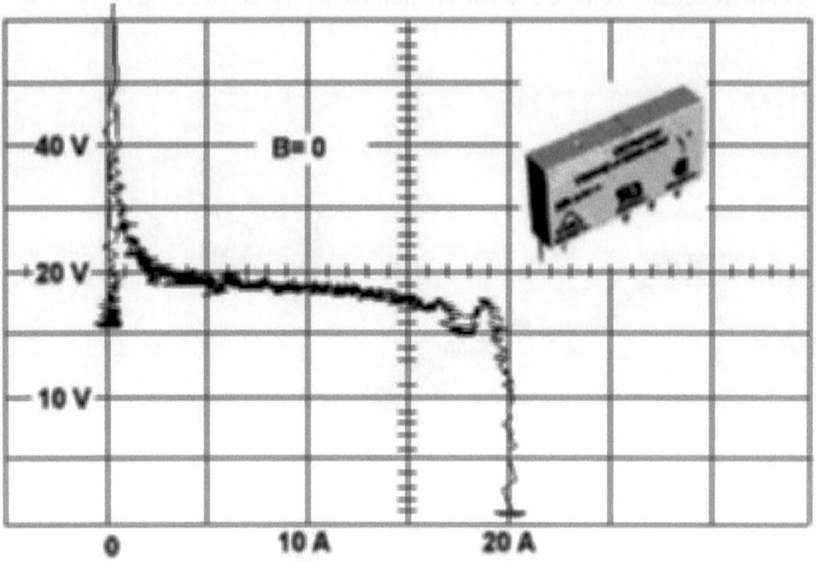

Fig. 7. SNR V I diagram inductive load 20 A, 10 mH B=O

In this experiment the arcing voltage for a given current at the highest magnetic field was up to 10 times higher, than at zero field. The increasing of the arc length cannot explain this tremendous voltage increase. The measured arc voltage at 10 A was 200 V. This voltage corresponds to an arc length of 110 mm [10]. The size of the relay is 5x14x28 mm³., so the whole relay had to be completely filled with the arc. Additional experiments under reduced air pressure showed, that this effect is strong dependent of the contact gap width.

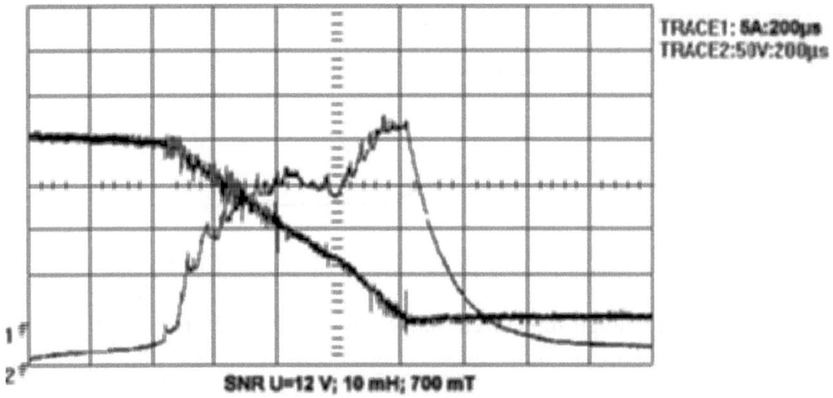

Fig. 8. SNR contact voltage and current inductive load 20 A; 10 mH B=O.7 T

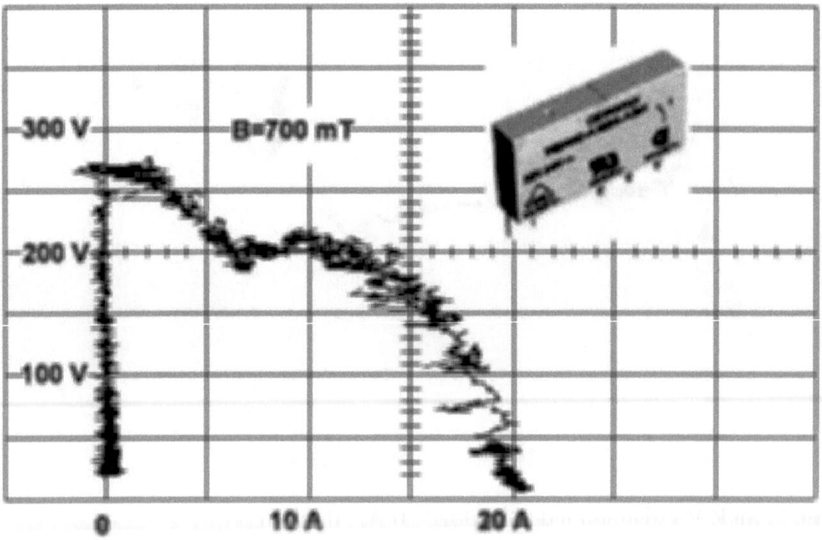

Fig. 9. SNR V I diagram inductive load 20 A; 10 mH B=O.7 T

2.4 Switching High Current Loads

Fig. 10. shows the breaking capability of relays with form U contacts, like the tyco Battery Disconnecting Switch (BDS) and the tyco High Current Relay (HCR). Both relays are available in a latching version. These relays were originally developed to switch high current loads (I>100 A) at 14 VDC. They will also be used as emergency interrupters.

Since we had no batteries for 42 VDC power net, we used three standard 12 V 36 Ah batteries in series. The internal resistance of this battery block was approximatly 30 mΩ. The current was set to 1000 A by a 6 mΩ external resistor. The voltage measured at the battery poles dropped to 10 VDC, when the load was switched into the circuit. At the moment the contacts separate, the contact voltage rises to 12 VDC. The current begins to decrease. The second contact opens approximately 200 µs after the first one, because the relay had a contact height offset. At this moment the current gradient of the current decline became much steeper. After 1.2 ms the current was completely turned off.

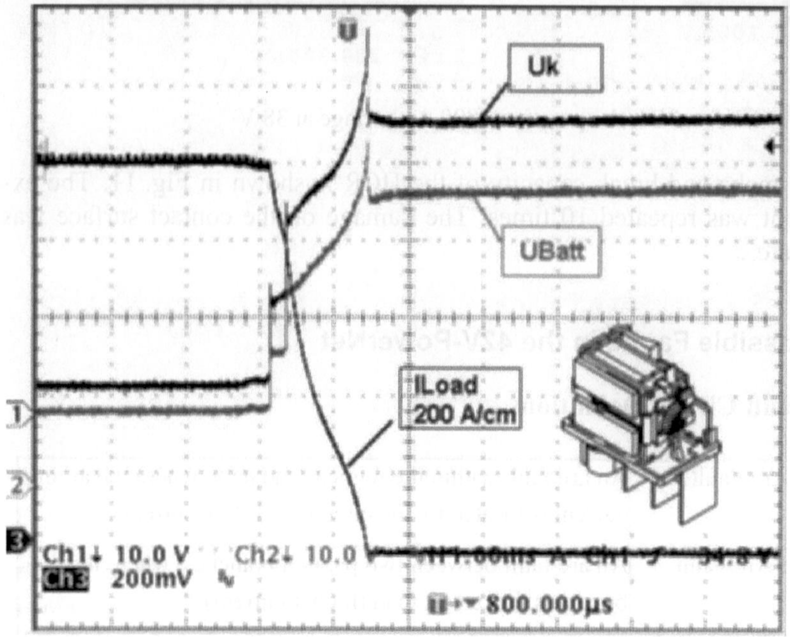

Fig. 10. HCR High Current Switch Off Capability

Besides the capability to switch off a short circuit, it is also necessary that the relay is able switch into a previously shorted circuit and break this connection successfully after the detection of the fault condition.

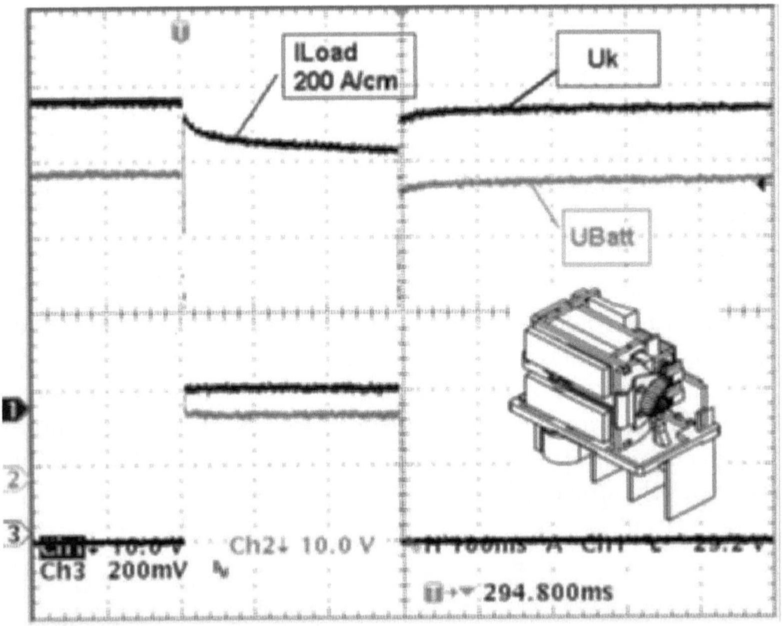

Fig. 11. Make and Break operation 1000 A shortage at 38 V

The make and break capacity of the HCR is shown in Fig. 11. The experiment was repeated 10 times. The damage on the contact surface was moderate.

2.5 Possible Faults in the 42V-PowerNet

Arc Fault Circuit Definitions

Series Arc Fault	An arc fault within the regular load path, where the arcing current is limited by the circuit load (low current)
Parallel Arc Fault	An arc fault between two parallel branches. where the arc bypasses the circuit load (higher current)
Ground Fault Arc	An arc fault between circuit and ground (including arcing to adjacent active lines)

only dual-voltage power net

Residual or leakage current	A fault current due to a current flow between the normally strictly separated 14 V and 42 V power-net

Compared to the situation at 14 VDC, the higher voltage will cause new serious faults in the power net. A dual voltage power net architecture with two coexisting voltage levels, i.e. 14 VDC and 42 VDC, will enhance the number of possible failures as well. Without additional measures, these faults will have serious consequences, because of the high energy dissipation in case of arcing or misguided currents.

2.5.1 Series Arc Fault

Series arcing is caused e.g. by a loose connection in series with the load circuit. Series arc current is limited to a moderate value by the resistance of the device that is connected to the circuit. The amount of energy dissipated in the sparks from series arcing is less than in the case of parallel arcing. But even a few amps are enough to be a fire hazard or lead to series arcing is, which particularly insidious because the arc current remains well below the rating of the fuses. Since the peak current is never higher than the steady-state load current, series arcing is much more difficult to detect than parallel arcing. Furthermore, there are also natural occurrences of arcing, for example when switches or relays are activated. These short-term arcing situations are generally not hazardous and should not cause nuisance tripping.

2.5.2 Parallel Arc Fault/Ground Fault Arc

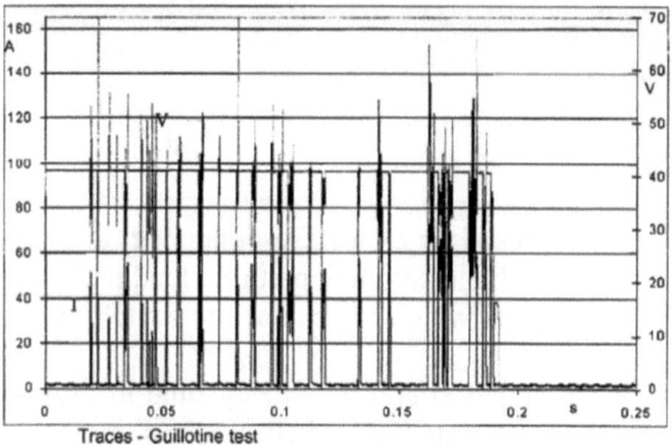

Fig. 12. Example of parallel arcing [11]

Parallel arcing occurs when there is a direct short circuit between two power wires and the current is limited only by the internal- resistance of the power supply in the distribution circuit.. As indicated in Fig. 12., the current could flow intermittently and the average current may not be sufficient, to be recognized by the fuse. On the short time scale, parallel arcing cannot be distinguished from full short circuit condition.

Parallel arcing is generally more hazardous than series arcing for two reasons: The energy in the sparks is much higher and the hot metal that is ejected is more likely to come in contact with inflammable material.

2.5.3 Sensor Principle for a Series Arc Detector

Arcing can occur at any location within the current path. The simplest way to notice that there is arcing might be to listen to the radio. The information which load path is affected, can not be determined from the radio signal. The arcing sensor described here is designed to detect series arcing. The only global available information about an arcing event in a particular load path can be found in the additional modulation of the load current, i.e. the arcing noise.

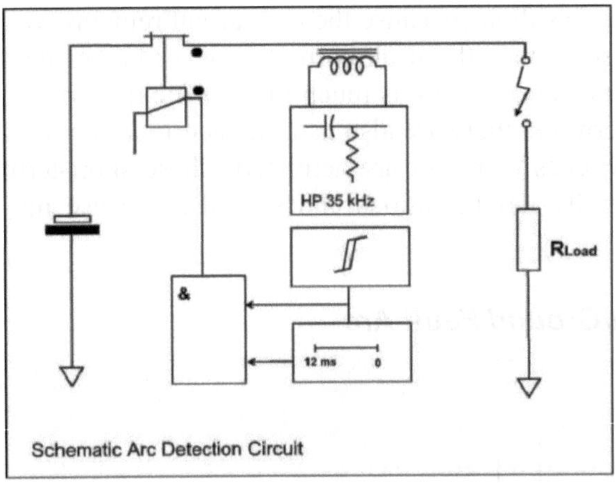

Fig. 13. Scheme of Arc Fault Detection Circuit

The arc characteristic noise is picked up by a RF oscillator set to a certain frequency, which is coupled with the common part of a load path. The detection circuit gains a trigger signal, which can be used to switch off the load path in question.

The trigger signal will be generated after 12 ms delay time to prevent that "normal" non hazardous arcing events (i.e. switches, relays, motors,..) would cause tripping.

Performance

Our standard test set up consists of an adjustable 60 VDC, 50 A power supply, various resistive and inductive loads and a single break relay used as an arcing generator.

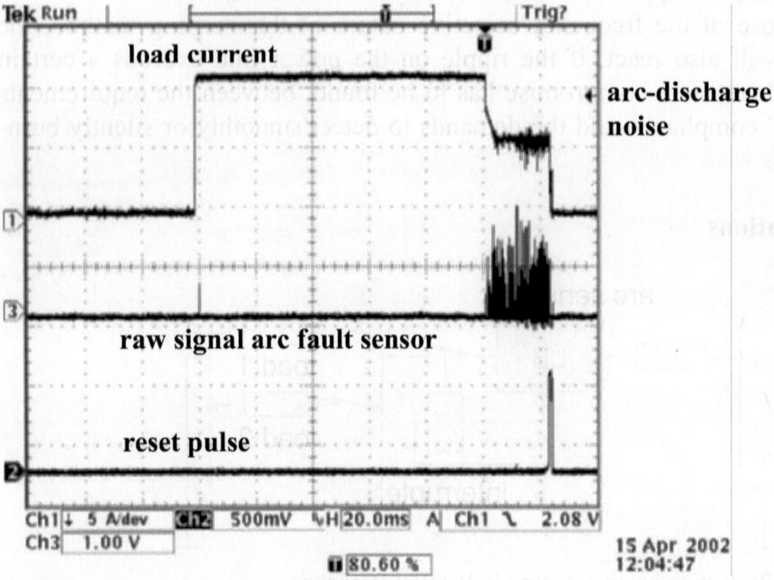

Fig. 14. Arc Detection Characteristic

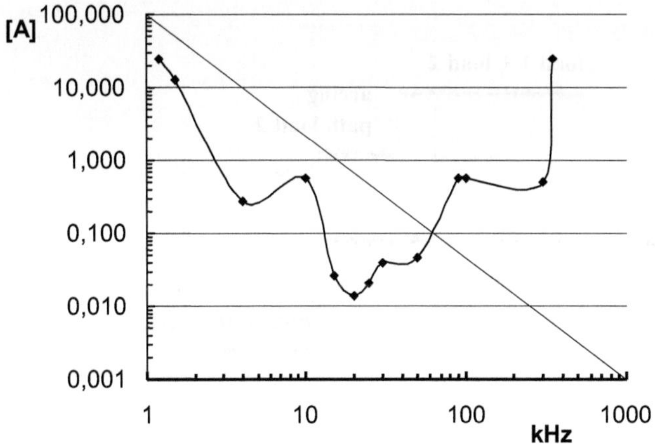

Fig. 15. Allowed ripple trip current vs. frequency

Trace 1 in Fig. 14. corresponds to the current, trace 3 contains the amplified arc noise signal behind the input filters and trace 2 depicts the trigger signal. The arcing is generated 100 ms after make, by breaking the single contact of the arcing generator. The arcing triggers a delay which blinds out normal switching events. As after 20 ms the arc is still burning, the interrupter tripped and the load path is switched off.

Because of the frequency-selective nature of this sensing method, the sensor will also react, if the ripple on the power line exceeds a certain limit. Therefore a compromise has to be found, between the requirements of EMC compliance and the demands to detect smoothly or silently burning arcs.

Applications

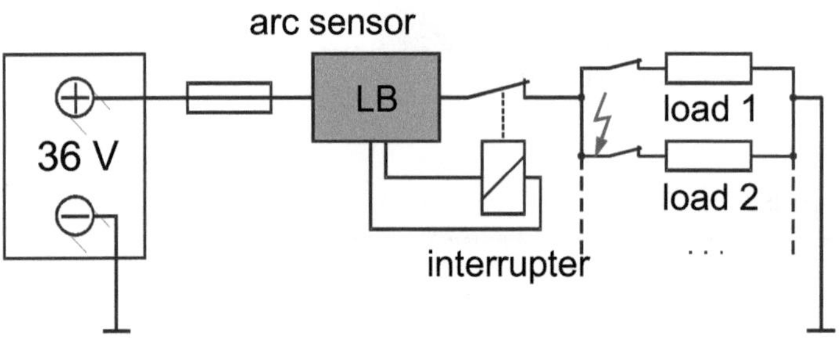

Fig. 16. Protection against Arcing in Parallel Branches

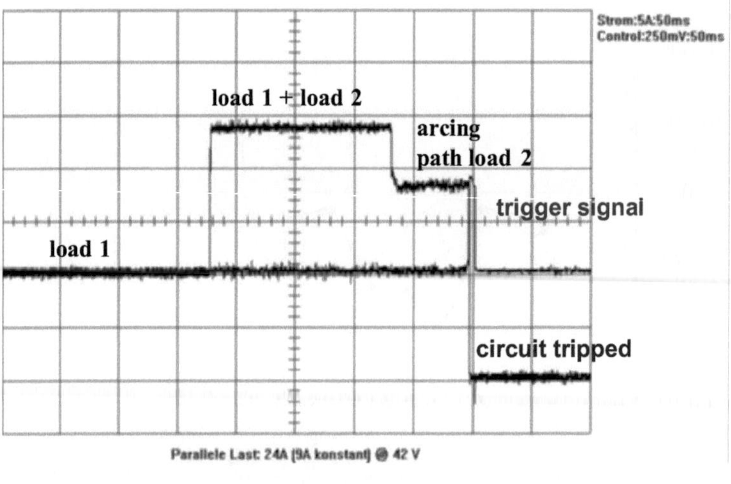

Fig. 17. Monitoring of Parallel Loads Arcing in one Load Path

By monitoring the current, every arcing event within the load path, will trigger the sensor, independent from further wiring. As the sensitivity of this sensor to detect arcing is more or less independent from the actual current load. It is possible to detect arcing for groups or clusters of loads in parallel branches.

The circuit reacts as soon as in one load path arcing occurs, independent of the current flow through the parallel loads.

As sketched in Fig. 16. the current trace in Fig. 17. corresponds to two parallel loads, which are switched subsequently into the monitored load path. Load 2 was switched 180 ms after load 1 into the circuit. Then an arc was drawn in the load path 2 and further 60 ms later the interrupter tripped. The trip delay in this example was set to 60 ms.

Application Hot Disconnect

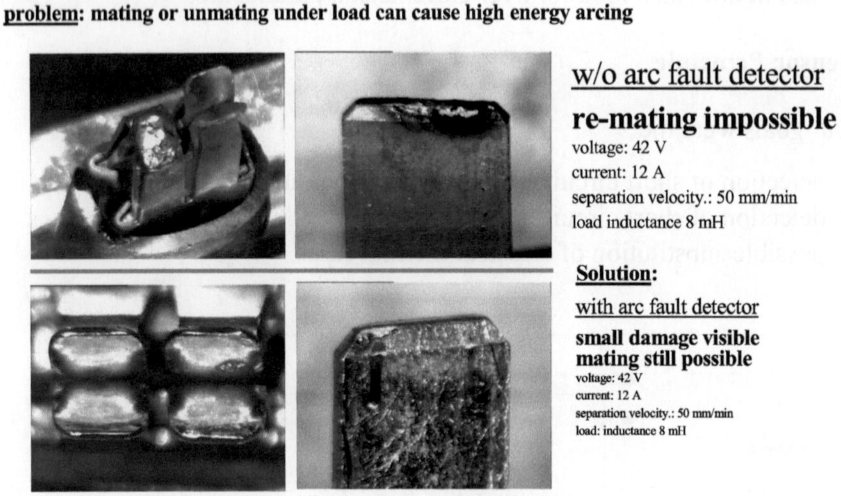

Fig. 18. Damage due to Drawn Arcs by Hot Disconnection

The series arc sensor is able to protect male and female connectors against serious damage. Because of the sensing method, which actually needs arcing to react, small visible marks on the connector surface cannot be avoided. Nevertheless the connectors remained connectable after repeated mating and disconnecting actions under resistive and inductive load. The results shown above were gained with a trip delay of 12 ms.

2.5.4 Leakage/Residual Current Fault

In case of a short circuit between the voltage levels in a dual-voltage power net, high currents could arise. Although it is very likely, that 14 VDC loads would be destroyed by the over voltage. If all 14 VDC loads were protected by a voltage limiting device, this device would have to absorb a lot of energy. If for instance the 14 VDC voltage level is clamped to 22 V, then with the internal resistance of the 42 V battery (e.g Ri = 100 mΩ)

$$\Delta U^2/Ri = (42V - 22\ V)^2/100\ m\Omega \qquad (2.)$$

the clamping device will have a power consumption of 4 kW. The time this device could carry such a burden would be limited. The rough estimation shows, that it might be useful to monitor such misguided cross currents. The sensor described below will not only detect these residual currents, but does also monitor over currents and parallel arcs.

Sensor Principle

Our goals were the

- detection of short circuits between the 14 V and 36 V voltage levels,
- detection of short circuits to ground and
- possible substitution of classical thermal fuses.

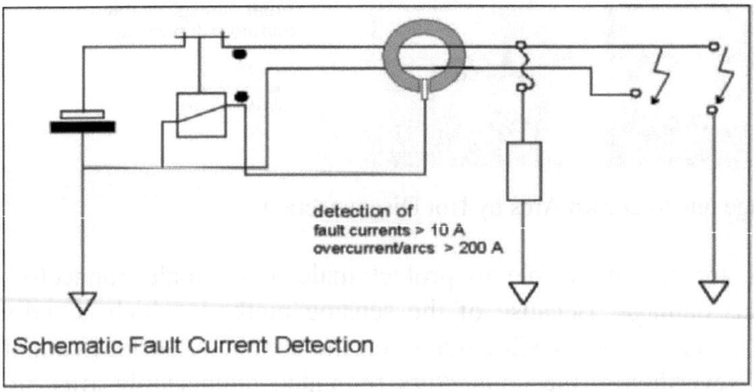

Fig 19 Differential current sensor principle

The current sensor consists of an iron ring, the yoke, with a half turn of the go and a half turn of the return line. The current flows antisymmetrically through this wire turns. Further a hall switch is placed in the gap of the yoke. The magnetic flux through the iron is then monitored by the hall switch. In the case of small equal currents ($I<100$ A) in go and return line, the magnetic flux in the yoke is nearly compensated. At a preset magnetic flux value, corresponding unbalanced currents in the go and return line, the hall switch toggles. The tripping level can be varied in wide ranges by an additional auxiliary magnet. The hall signal then will reset a latching relay. Because the yoke-hall-switch unit is not perfectly symmetric and additional saturation effects within the yoke, the sensor will also trip, if the balanced currents exceed 200 A. Therewith the leakage current sensor can be used as an over current detector, for parallel currents higher than 200 A.

2.5.5 Performance of the Interrupter

Assuming the above described faults in the 42 VDC power net are detectable, what kind of switching element has to be used to disconnect the supply line in case of parallel arcing? The peak current is only limited by the internal resistance of the power supply. Even if the avarage current within a parallel arcing event is not sufficient to trip a conventional fuse, the leakage current sensor will generate a switch off signal. But the sensor can not distinguish a hard short from parallel arcing. Therefore the switching element has to be capable of disconnecting the maximum expected current rating, i.e. $I_{max}=U_{Batt}/R_i$.

2.5.6 Switching Capability of the tyco µK Relay

For a protection device the performance of the detectors are as important as the capability of the switching element. These elements should have all the advantages of conventional fuses, like minimal size, low power consumption and they have to keep their state after the tripping event occurred. We chose the tyco µK Relay. This 30 A rated relay has the size of a sugar lump. With an auxiliary magnet attached to the contact chamber, this originally 14 VDC breaking relay was upgraded for 42 VDC applications.
This type of relay is available in mono-stable and latching version. Using the latching principle for the protection device has the advantage, that this relay under normal conditions has no power consumption. Also the response time of a latching relay is faster because of the, compared to mono-stable relays, reduced influence of the coil inductance.

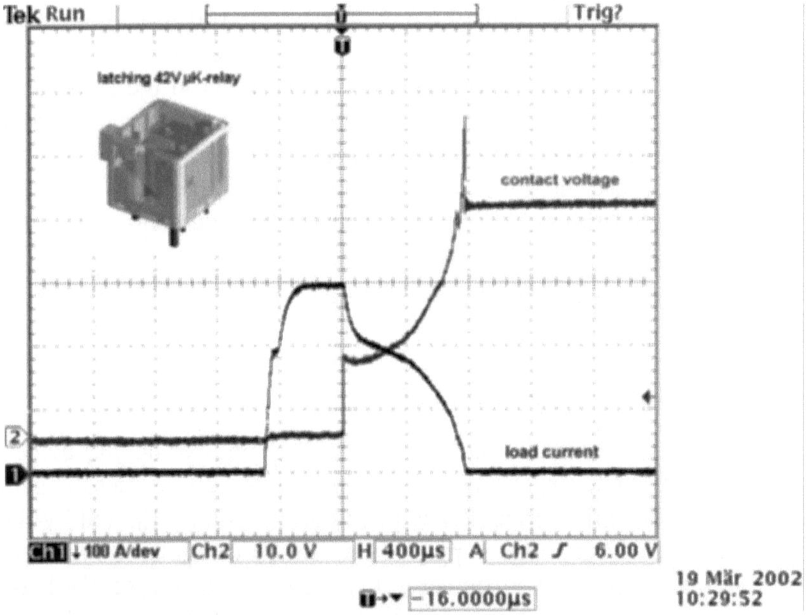

Fig. 20. Disconnection of 300 A Fault Current

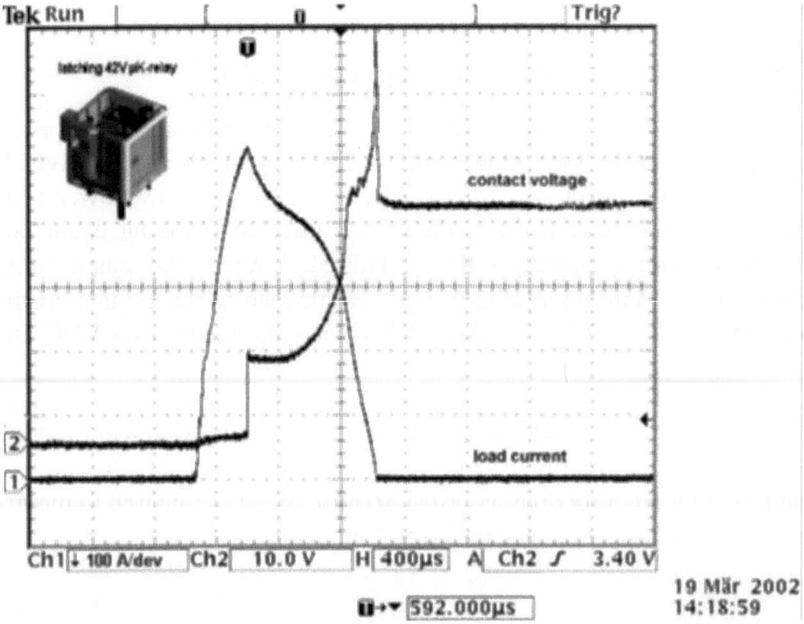

Fig. 21. Disconnection of 500 A Fault Current

The current rises up to 300 A within 160 μs (Trace 1). The effective circuit inductance was estimated as 10 μH. The first discontinuity after 80 μs was identified as bouncing of the auxiliary HCR relay. After 600 μs the contact opens and the contact voltage rises stepwise to 16 VDC. Then during the contact movement the arc burns for 800 μs with rising arc voltage until shut down.

In case of a ground fault arc, the peak current is only limited by the internal resistance of the power supply. Therefore, if the protection device recognize this peak as an error condition, the switching elements have to be capable to deal with the high current, even if it is for a short period. The reaction time of the detector switch unit has to be kept short. The response time is mainly given by the relay coil inductance and the inertia of the armature mass. The used latching type relay has the advantage to neglect this time delay caused by the coil inductance. If an over voltage is applied on the reset terminals the relay toggles immediately. We have switched off 520 A multiple times, with one relay, but failed to switch higher currents. Not that we have reached the relay limits, but the reduction of the load resistor seemed to have no effect on maximum current.

Obviously the relay opens faster in the 500 A test (400 μs), than in the 300 A test (600 μs). Due to the small inductance the current could not reach it's steady value within switching time.

The high current generates an extra repulsive forces within the relay, which leads to a faster response.

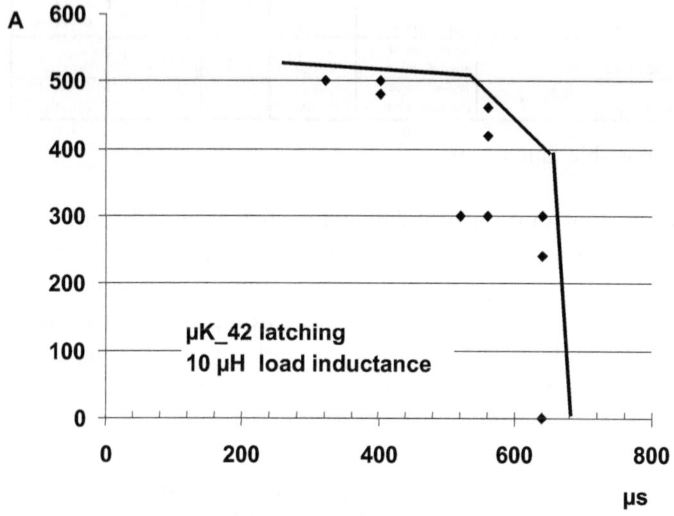

Fig. 22. Current vs. break time

At first, we calculated the electrodynamic repulsion force in the symmetric contact sphere [9] and found that this contribution is only in the order of 1/10 of normal contact force. Obviously there were other contributions. Then we calculated the dynamic repulsion force, due to the inductive coupling of the load path and the armature. The high dI/dt in the load path generates eddy currents in the armature. The resulting repulsion force reduces the effective contact force to 1/2 of it's normal value.

The relay response time in the low current limit (up to 300 A) is mainly determined by the inertia of the armature mass. Fig. 22. indicates, that for the given circuit inductance of 10 μH, the relay will limit the maximum current at 600 A by itself independent of the external load.

2.6 Conclusion

It is possible to recognizes arc faults and misguided currents with the presented detection principles. The matrix below summarize the presented results.

Matrix of possible faults				**protection device**		
short circuit or arcing event (assumed Ri(12V) = 12 m Ω ; Ri(36V)=100 m Ω)		**remarks**	**est. Curr. [A]**	**fuse**	**leak. curr. Detector**	**series arc sense**
plus 42 V	thru 12 V load in the case 12 V battery is discon.	direct	0 - 200	-	> 5 A	**detected**
plus 42 V	*thru 12 V battery*	direct	200	*12V/42V*	detected	**detected**
plus 42 V	*ground*	direct	400	*42V*	detected	**< 10 ms**
42 V load with lost ground	thru 12 V load in the case 12 V battery is discon.	lost ground	0 - 100	-	> 5 A	**detected**
42 V load with lost ground	*thru 12 V battery*	lost ground	0 - 60	*12V*	> 5 A	**detected**
plus 42 V	*thru 42 V load*	*arcing due to disconnection*	0 - 100	-	-	detected
42 V load with lost ground	*42 V ground*	*arcing due to disconnection*	0 - 100	-	-	detected

Fig. 23. Matrix of possible faults

We have shown, that our sensors can detect the assumed fault conditions. The interrupter will securely separate the affected load path when triggered. Parallel arcing could be detected by the over/leakage current sensor, whereas the arcing sensor only is a protection against series arcing. Both trigger signals could be used to reset a latching relay from 1 cubic centimeter size. If the reaction time is short, then these normally 30 A rated relays are capable to switch 500 A. We could show in our experiments, that if the relay gets the reset pulse in time, the load path will be opened under all circumstances. As a surprising advantage, these relays showed in the experiments with the 10 μH circuit inductance a current limiting behavior. This short circuit capability, normally the purpose of conventional fuses, opens the way for a remote controlled fuse. This device could not only monitor arcing events. With a few insignificant changes, It might be possible to use the detector relay unit for wire protection purposes.

3 References

[1] Rieder W, Physikalische Grundlagen elektrischer Schaltgeräte, Vorlesungsskriptum, Wien, 1995

[2] Keil A, Merl W, Vinaricky E, Elektrische Kontakte und ihre Werkstoffe. 2. Aufl., Springer – Verlag, Berlin 1984

[3] Rieder W, Low Current Arc Modes of Short Length and Time IEEE Transaction on CPT Vol 23 No 2, June 2000

[4] Rieder W, Plasma und Lichtbogen, Vieweg, Braunschweig, 1967

[5] Erk A; Schmelz M, Grundlagen der Schaltgerätetechnik, Springer, Berlin, 1974

[6] Ayrton H, The Electric Arc, The Electrician Comp., London, 1902

[7] Slade P, Electrical Contacts-Principle and Applications, Marcel Dekker, New York Basel, 1999

[8] Schoepf T, Heinrich J, Breaking Capacity of Automotive Relays and Switches in 42 VDC Power Networks, Proc. 48th International Relay Conference, Lake Buena Vista, 2000

[9] Holm R, Electric Contacts, Hugo Gebers Förlag, Stockholm, 1946

[10]Rieder W, Die Stabilität geshunteter Gleichstromlichtbögen, ELIN-Z VII Heft 3, Wien, 1955

[11]Hetzmannseder E; Zuercher J, 42V Arc Faults:Physics and Test Methods, Proc. SCIWORX Forum Vehicle Electrical Architechture, Hannover, 2001

Detection and Characterization of Short Circuits in 42 V Power Nets

Reinhard Seyer, Roland Fischer, Rainer Mäckel

DaimlerChrysler, Frankfurt

Horst Brinkmeyer, Fritz Schmidt, Thomas Schulz

DaimlerChrysler, Sindelfingen

Abstract

There is a qualitative difference between the 14, 28 and 42VPowerNets, which is related to their protection against faults due to the generation of arcs. The reason for that is the different behavior in the case of arcing according to the different voltages. While arcing occurs as serial and parallel arcs, this paper concentrates on parallel arcing caused by short circuits. It describes safety-critical relations, simulation, modeling and attempts to standardize the classification. Finally a protection concept is described and some examples of measures to solve the problems are given. In particular this paper discusses an attempt to standardize and to get a reference for testing the quality of protection components and methods and to stimulate further investigations.

1 Introduction

14 V and 28 V are the generally applied voltages for the automobile power nets in passenger vehicles and trucks, respectively. To account for the growing need of electrical power in automobiles, the power net of 42 V is under intensive investigation and test. The subject of this paper is the formation of short circuits and the protection against them. The main problem is, that they are associated with arcing. The arcing problem increases with

voltage and becomes essential in power nets above a certain voltage level. The voltage is responsible for arcing, because there is a need of a certain cathode and anode voltage drop in order to sustain electrical arcs. The critical value is about 10 V at each electrode. This voltage drop at the electrode is the reason, why there appears to be no problem in the existing 14 V power distribution systems, there might be problems in 28 V networks and real problems are expected in 42V-PowerNets.

2 Short Circuits in 42V-PowerNets

For short circuits, that are not associated with arcing, there is a well-established solution: In this case cables and components can be protected by the specified fuses. Investigations with several prototypes did practically present no problem. These short circuits are mainly caused by low-resistance contacts, that happen to be shorted in the unpowered state.

When, however, the short circuit is associated with arcing, the behavior is completely different.

The conditions for the occurrence of arcing are given in an electrical system above a certain voltage level. It can be distinguished between serial and parallel arcing. Serial arcing might occur by mechanical switching, by changing fuses and by any other interruption of the continuous current flow. Fortunately there is no problem with semiconductor switches. Possible solutions are dedicated electro-mechanical relays, prohibiting the change of fuses under power and respective safety hints.

Parallel arcing occurs in connection with short circuits produced by the contact of a power distribution cable with the chassis. Possible reference to a realistic event could be given by cutting into a cable by an edge of the body in the case of a crash. Another scenario might be given by degradation of the insulation of a cable in case of stress caused by continuous squeezing of the insulation in combination with vibration. In both cases there will be a connection from the high side of the network to ground. If this happens in the case of a powered system of 42 V, the short circuit will in nearly all cases be combined with arcing. A picture of the reaction can be seen in Fig. 1.

In this case the short circuit energy is focussed to a small spot, that leads to an extremely fast increase of the temperature, arcing, melting of the surrounding materials and explosions, which distribute the molten particles of copper and steel in the area. The arc is, however, not stable, it is fluctuating and stays as long as there is enough material to support the short circuit. The fluctuation of the arc is represented in a corresponding behavior of the current. A typical curve of its fluctuation is shown in Fig. 2.

Fig. 1. Arcing associated with a short circuit

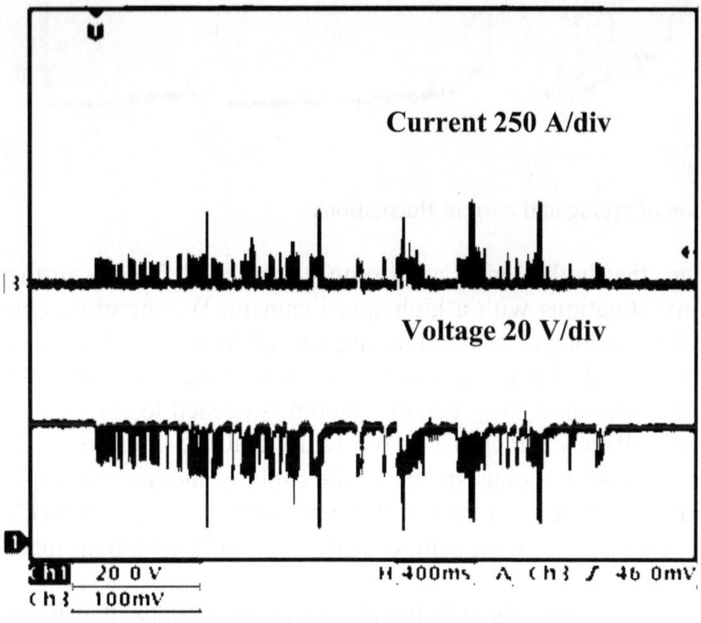

Fig. 2. Fluctuation of the current in case of arcing

We found that the fluctuation of the current is closely correlated to the arcing process. To establish this correlation, we have made a simple experiment: A photo diode was placed close to the expected short circuit and its signal was synchronized with the measurement of the current. Its result can be found in Fig. 3.

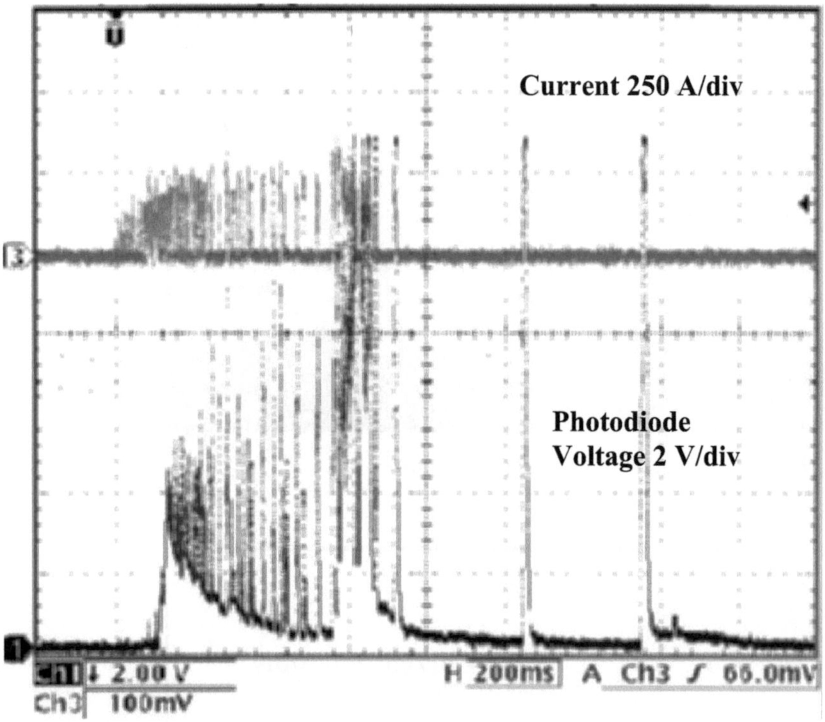

Fig. 3. Correlation of arcing and current fluctuations

It can be seen, that both signals are strongly correlated. This is further supported by investigations with a high-speed camera. We therefore conclude, that the fluctuations of the current are caused by arcing. The arc is not stable and we found, that the reason for that is twofold. First, for a stable arc an internal resistance of a certain amount is needed to stabilize its negative current-voltage characteristic. We found that the value is in the range of 0,5 Ω. Second, the consistency of the cable is another reason for the instability of the arc. It is, of course, not made of bulk copper, but consists of several strands. We could achieve stable arcs only with bulk material.

In the case of stable arcs, there is the dangerous possibility, that it migrates in a bundle of cables or between a cable and the chassis, if they are attached in parallel. We made experiments to this and found in no case arc migration. The arc is too instable, and the insulation is too thick as that migration of the arc could occur. It extinguishes itself as soon as the corresponding material is consumed.

The most important disadvantage of arcing results is the fact, that the corresponding average current is restricted by the fluctuations. And besides that, it is restricted by the voltage drop of about 20 V and by the positive thermal characteristic of the accompanying conductor materials. The problem is that a related fuse will not blow in any case. The average current is reduced by the arc and its reaction is no longer strong enough to melt the fuse. This leads to the essential problem of danger of ignition, which will be dealt with in the following chapter.

3 Experimental Approach

With three different experimental setups we tried to simulate realistic short circuits. The first was a cutter activated by hand as depicted in Fig. 4.

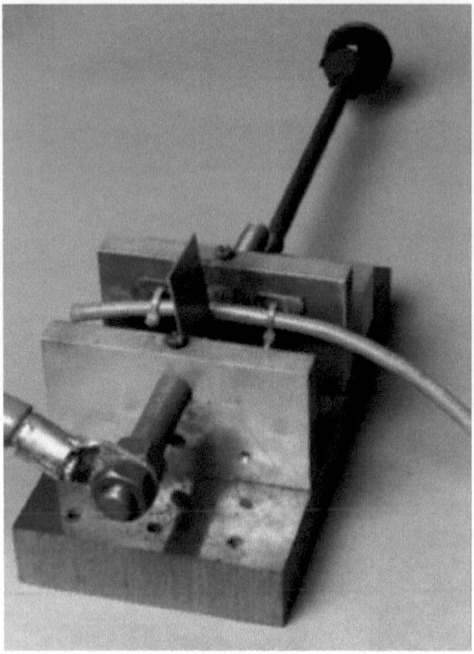

Fig. 4. Cutter activated by hand

With the tests with this setup we wanted to find out what happens if there is a slow penetration of the steel into the insulation and the copper strands. The second machine looked like a guillotine. A steel blade connected to the guillotine cuts very fast into a cable. A picture of it is given in Fig. 5.

Fig. 5. The guillotine-like test setup

Fig. 6. Saw-like test equipment

With these experiments we simulated fast penetrations of the body steel sheets into the copper cables in the case of a crash. The last simulation was realized by a saw-like machine. In this case a motor was connected to a steel blade. The blade was moved back and forth like a saw, until the insulation was abraded, and it came to an electrical contact and a short circuit. The realistic event simulated by this setup is the slow wear of the cable insulation by a sharp steel edge. A picture of it is given in Fig. 6.

With the saw we produced the most critical results. In general it can be said, that the phenomenon of arcing is connected to the time scale of the short circuit. High-speed motion results in fast intrusion of the steel into the copper, the contact resistance decreases very fast to a very low value. The result is a very high current. However the power consumption in the short circuit point is limited by the low contact resistance. Due to this in most cases the corresponding fuse was blown. In contrast to that, using the saw the resistance is high, and although the current is reduced, there is an increase of energy. We obtained most results under worst-case conditions by using the saw.

We investigated the reaction of fuses in combination with their corresponding cables. And we found that there is no problem with fuses in the range of 5 to 15 A. But from 30 A on fuses did not blow in any case.

4 Characterization of the Criticality

The behavior of the fuses can be explained very simple. The current is not high enough to support their melting integral. In fact, fuses were introduced to protect cables, and that is what they do. In case of arcing the cables are not in danger, because the current is not high enough. The critical point is, that the energy evoked by the short circuit is concentrated on the very small area around the contact.

Concentrating the energy in this small area results in very high temperatures and the question is, which materials might be ignited by this temperature. Flammability, ignition temperature and energy characterize these particular materials. Criticality is based on these inputs: the generated energy and the materials, that can be found in the neighborhood of the cables.

In order to define a realistic way to characterize, whether a short circuit could be seen as critical or not, we had to bring these two inputs together. We studied materials, that are integrated in the trunk and found, that the most critical one, that could be ignited with the lowest energy, was rubber. It was closely followed by foam plastics. It was found, that also the insulation can be ignited, but it did not burn self-supporting and the corresponding energy was higher than for foam plastics.

The grouping of the materials according to flammability has been performed by wrapping around the cable a stripe of the material under test to expose it to the energy generated by the arc, and than sawing through it.

The quantification of the related energy has been studied in more detail in combination with foam plastics. We used a cable of 10 mm² and small stripes of the test material of 15 cm length and 4 cm width. It was found, that energies below 500 Ws do not ignite the plastic. Energies above 600 Ws lead in nearly any case to an ignition. Between these values there is an undefined area.

The results can be seen in Fig. 7.

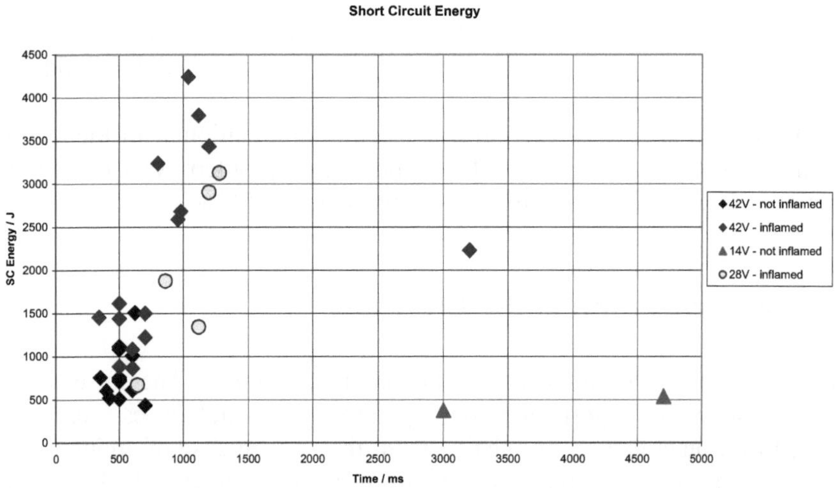

Fig. 7. Ignition characteristic of foam plastics

Two things are important. The area below 500 Ws is largely uncritical, and all energies above this level are critical. A corresponding curve can be derived for all respective materials. There is not only a dependence on the energy, but the time scale also plays a role. If the energy is spread too widely in time, the cooling effect of the cable will reduce the temperature and no ignition will occur.

5 Thermal Model

The physical effect is determined by the energy supply from the short circuit and the consumption of energy by the melting of copper and steel, radiation, and the cooling effect by the conduction of heat through the cable. It was found that, for a specified situation, the melting takes about 30 % of the energy. Radiation was estimated to be in the range of 10 %. To determine the heat transfer by the cable, we made some investigations with a temperature sensor placed near the short circuit. Fig. 8. depicts the temperature in the cable in dependence of the time.

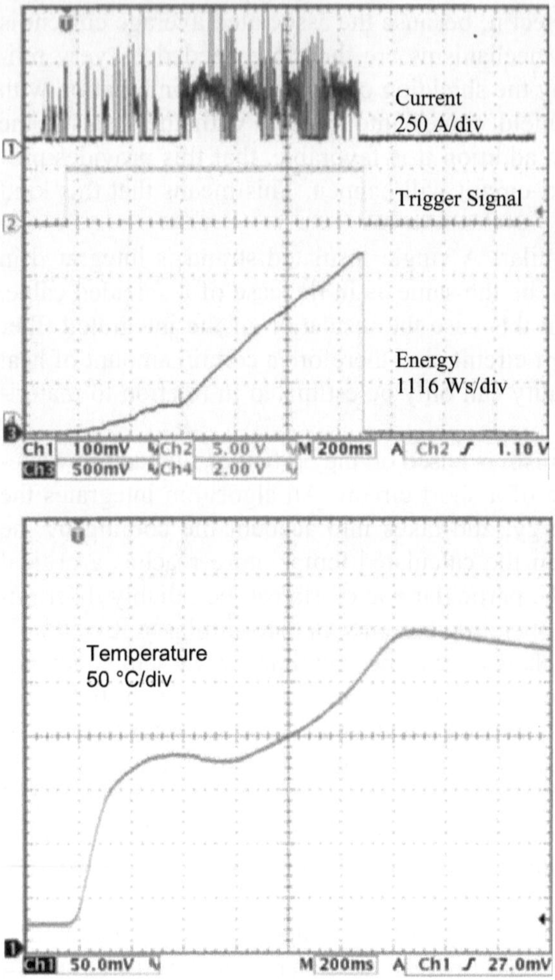

Fig. 8. Behavior of the temperature in the cable near the short circuit

It can be seen, that there is a rapid increase and that it depends on the supported energy. With these measurements we found, that the heat transfer in the cable takes roughly 60% of the supported energy. The thermal model can serve as a useful object of further investigation, because a realistic model can be a very good reference for testing and qualification of various arc detection mechanisms.

6 Detection of Short Circuits and Protection Concept

The detection of short circuits combined with arcing presents the real challenge. The fuse will not detect it, because the associated average current is too low. Other protection mechanisms are therefore needed. A very reliable one can be achieved by the shielding of the cable in combination with an active switch. If the shield comes into contact with the chassis, the switch will be activated. In addition it is favorable, that this provides protection before the real short circuit will happen. This means that this kind of protection is not critical.

A simpler method is similar: A single insulated strand is integrated in the cable. The reaction will be the same as in the case of a shielded cable. The disadvantage is, that in this case the insulation of the integrated fiber must melt to detect the short circuit, and therefore a certain amount of heat has to be produced. Criticality can only be estimated in relation to materials near the cable.

A third detection mechanism is based on the particular fluctuating waveform of the current in case of a short circuit. An algorithm integrates the power, determines the energy, and takes into account the cooling by the cable. It switches off, when the calculated temperature reaches a critical value. It is essential, that the particular waveform can be reliably discriminated and that there are no operating currents that are similar to it.

A fourth protection mechanism can be combined with semiconductor switches. If there is a current peak, that exceeds a certain level, the semiconductor will switch off. It will stay switched off for a certain time e.g. 4 ms and then switches on again. The assumption is, that in the time while the current is switched off, the melted material in the contact point will weld together and build a contact with very low resistance. If it switches on again, the low resistance will lead to a high current, that will be detected as a peak and it will be switched off again.

For a comprehensive concept it is an advantage, that for cables with a small cross-section conventional fuses are sufficient. Most of the cases can be covered by conventional means. Particular care has to be taken for cables of higher cross-section. These can be protected either by semiconductor switches or by switches in combination with a shielded cable or an integrated fiber.

7 References

[1] Seeliger R, mDie Bogenentladung, Handbuch der Exp. Physik, 13, S. 680 ff, 1929
[2] Kupfmüller K, Theoretische Elektrotechnik, S. 212 – 214, 10. Auflage,1973
[3] Phillipow E, Taschenbuch der Elektrotechnik, Band 5, S. 624 ff, 1981

Investigations of Electrical Failures in the Dual Voltage PowerNet

Ronald Große

Forschungsgesellschaft Kraftfahrwesen mbH, Aachen

Christian Amsel

Institut für Kraftfahrwesen, RWTH Aachen

Abstract

The Institut für Kraftfahrwesen (ika) of the Aachen University and the Forschungsgesellschaft Kraftfahrwesen Aachen (fka) have developed a test bench for the investigation of vehicle powernets. This test bench may be used to reproduce failures such as short circuits or electric arcing in current and future dual 42V/14V vehicle powernets. With this, one is able to predict the potential risks of failures occurring in the vehicle power-net, already in the early phase of development. Particularly new types of failures that could occur due to the planned introduction of an additional 42V voltage level can be investigated in this way without the need for a real test vehicle.

For the planning and evaluation of experiments, methodical approaches such as Design of Experiments (DOE) are used. As a result, the number of experiments are clearly reduced along with the possibility of easily applying the results to other situations. The individual steps in the investigation of failures in dual voltage powernets are described in this report.

1 Introduction

Due to the planned introduction of a 42 V voltage level in the electrical system of modern vehicles, apart from high costs and problems of availability of components there is also the potential for new failures in the electrical systems. In order to investigate these failures, test bench investigations are carried out as a first step. The advantage of test benches are the high flexibility and good reproducibility at defined boundary conditions. The number of possible combinations of experiments is limited by the respective design of the real experimental vehicle. Apart from that, the integration of the necessary measuring equipment into vehicles is also associated with higher expenditure. However, the limitations arising out of the abstraction of the vehicle onto a test bench have to be considered.

2 Concept of Investigations of Electrical faults in Vehicle Electrical Systems

2.1 Design of Experiments

Experimental methods are widely used in research as well as in industrial environments, in some cases for very different purposes. The primary goal in the investigation of electrical failures in vehicle powernets is usually to show the significance of an effect that a particular factor exerts on the desired variable.

2.1.1 Preparation of the Experiment

DOE is associated with the systematic planning and evaluation of experiments and hence combines specialized technical and statistical aspects.

The investigation of a failure and its effects usually requires more time than the execution of the planned experiments. The systematic approach acquired by the employees at the ika and fka proved to be suitable in this case. The individual steps of this approach are presented in Fig. 1.

In the first step, an attempt is made to define the problem in terms of a systems analysis. The result of this phase is a precise written formulation of tasks, the phenomena to be investigated and the desired parameters to be determined. For this purpose the system is examined by a group of specialists for diverse criteria, in order to identify substantial parameters, which could have an influence on the test results. Furthermore it is necessary to compute an experimental plan to determine the interacting effects between the parameters and the desired functions. These interacting effects could be differing in nature.

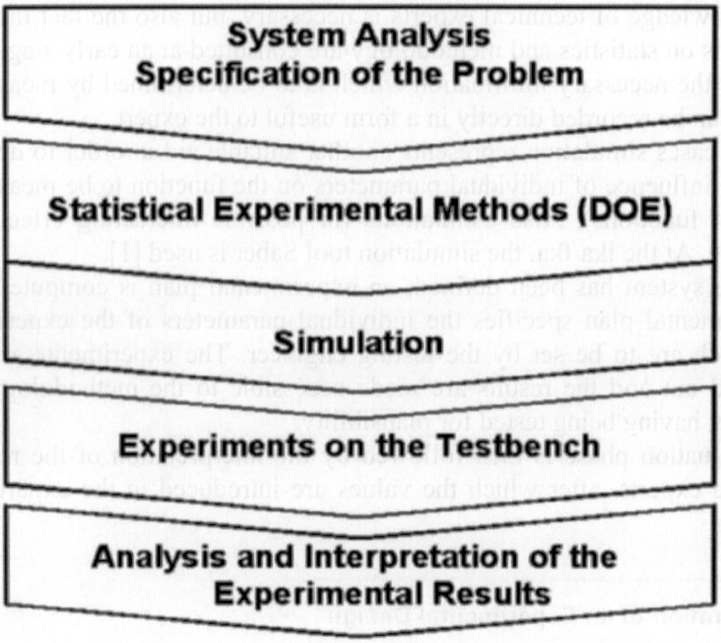

Fig. 1. Methodology

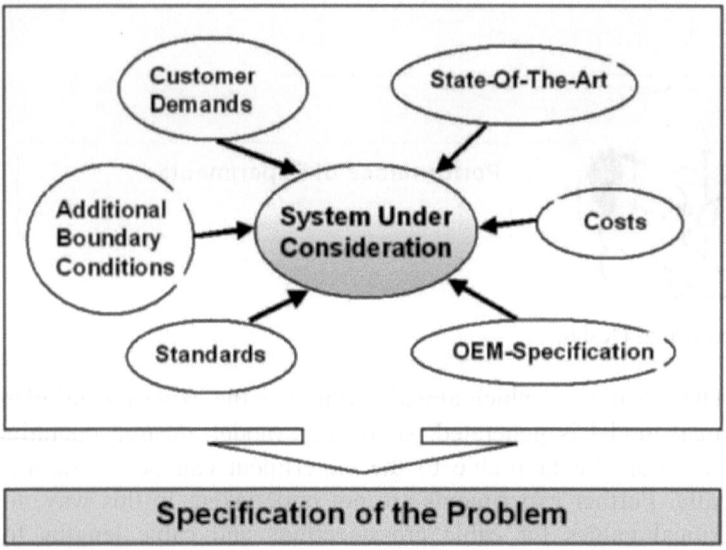

Fig. 2. System Analysis

In the context of the Design of Experiments however, not only the specialized knowledge of technical experts is necessary, but also the fact that these experts on statistics and methodology are consulted at an early stage. In this way the necessary information which is to be determined by measurements, can be recorded directly in a form useful to the expert.

In many cases simulation represents another suitable aid in order to determine the influence of individual parameters on the function to be measured (target functions). Also estimations for possible interacting effects can be made. At the ika/fka, the simulation tool Saber is used [1].

Once the system has been defined, an experimental plan is computed. The experimental plan specifies the individual parameters of the experiments, which are to be set by the testing engineer. The experiments are then carried out and the results are made accessible to the methodology experts after having being tested for plausibility.

This evaluation phase is then followed by the interpretation of the results by the experts, after which the values are introduced in the experimental plan.

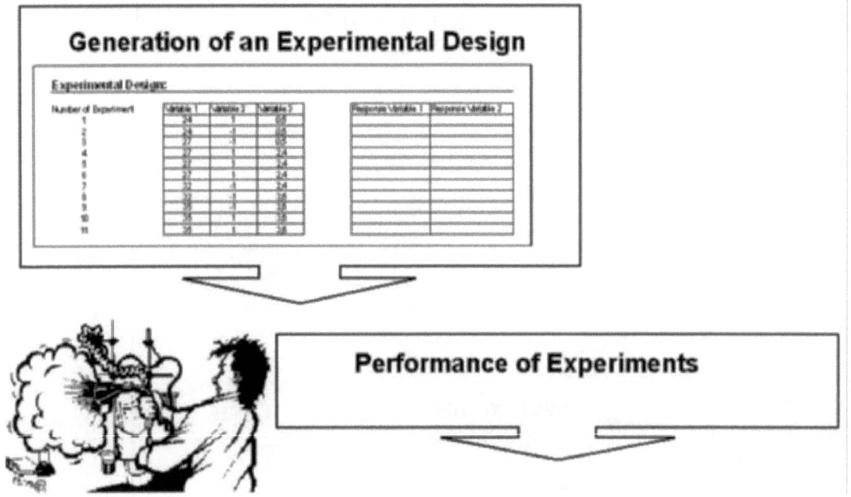

Fig. 3. Experimental Design

Using a DOE–Software which already computes the experimental plan, a mathematical model is generated. Using this model, desired operating points which would be favorable to the experiment can be pre-defined mathematically. Further experiments are not undertaken. In this way, for example, critical values for cable cross-sections and cable lengths for 42 V/14 V short circuits can be determined.

2.2 Components of the Test Bench

The test bench for the investigation of failures in vehicle powernets, is made up of several sub-systems. These sub-systems, for one, are made of components which are necessary in order to represent the powernet which is to be investigated.

2.2.1 Measuring Equipment

A high-quality measuring system is used for the measurement of various desired parameters. Using this PC-based system, both long-term measurements as well as short-term measurements with a time-specific resolution of 5MS/s can be carried out. All measurements can be carried out using potentially isolated channels, so that the detection of potential differences at grounding points across the vehicle body or measurements in a galvanically separated system are possible.

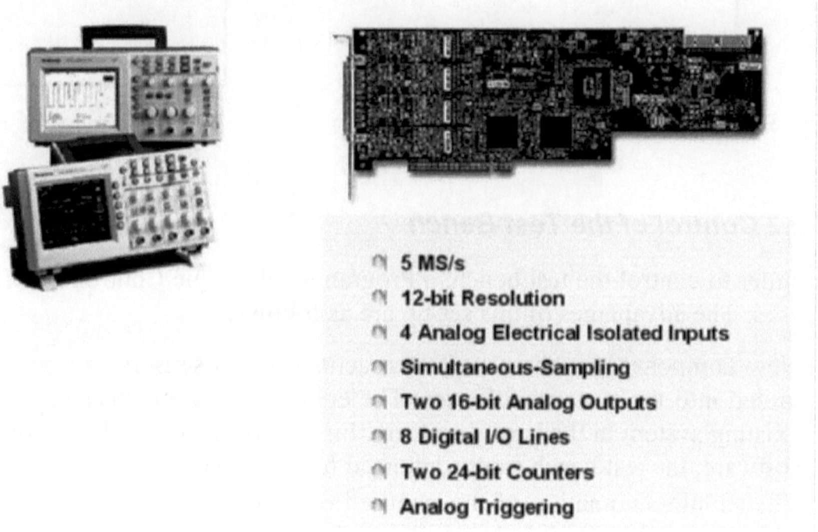

Fig. 4. Measuring equipment

The principle of measurement is described briefly on the basis of an example of current measurement: For the measurement of currents up to 2000 A, LEM transducers are used. Using these devices which function on the Hall principle, it is possible to fulfill the high demands on measuring speed, which is particularly required in the measurement of short-circuits and arcs. The measured signals, which are either generated in the form of voltages or currents, are then transferred to the input of a PC measuring card, where it is processed, saved and evaluated at a later point in time. Fig. 5. shows the principle of current measurement in the form of a block diagram.

Voltage measurements are carried out in an identical form and method and can be carried out with high precision.

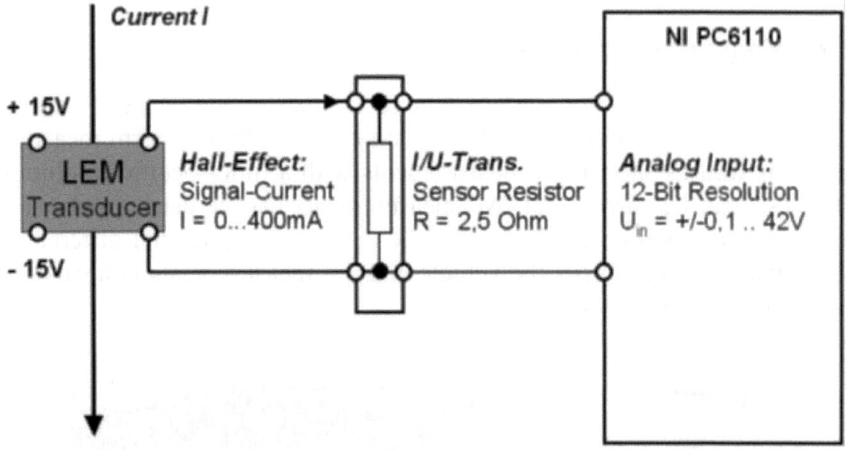

Fig. 5. Measurement of Current

2.2.2 Control of the Test Bench

In order to control the test bench, a Programmable Logic Controller (PLC) is used. The advantages of this set-up are as follows:

- New components such as switches, actuators and sensors can be integrated into the system with ease. The components are connected to the existing system in the hardware form. Finally, by a simple change of the software, the test bench can be adjusted for a new task.
- The simple automation of the required time-controlled switching processes can also be achieved without much effort. This results in better reproducibility of experiments as compared to manually controlled processes.
- Monitoring functions, such as the deactivation of components which exceed a pre-defined voltage level can be realised easily using a PLC and its analogue input ports.

The control of the test bench is explained as follows.

A control board with integrated switches for the operation of individual energy sources, storage devices and consumers, represents the user interface of the test bench. The control board can also be used for the reproducible generation of failures, i.e. relays for the production of short-circuits or units for the production of arcs can be operated.

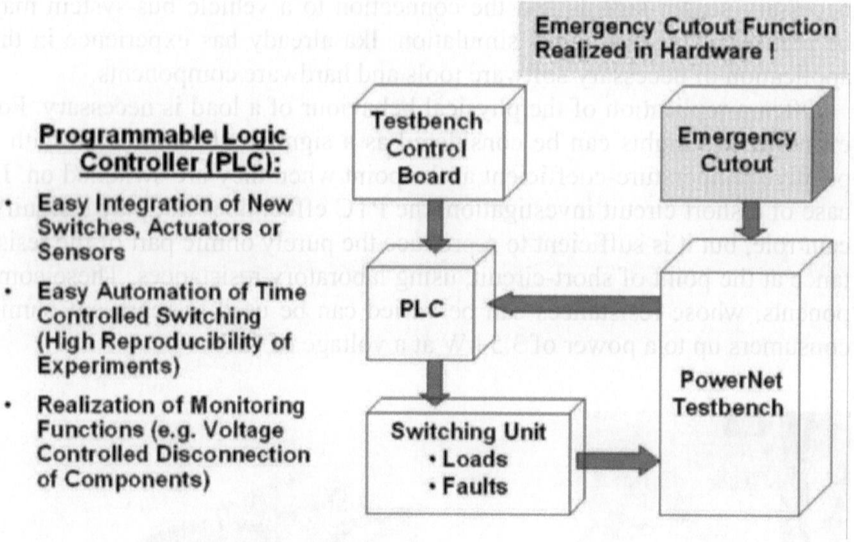

Fig. 6. Control of the Test Bench

All relays are protected by a hardware emergency-shutdown switch. In this way the VDE-standards for the operation of such equipment could be fulfilled. The emergency-shutdown switch is mandatory for fire protection, since the investigation of short-circuits can lead to uncontrollable cable fires.

The outputs of the PLC are used for switching the relays, which serve to control the different components of the test bench. In this way the power supply (alternator and batteries or supercaps) and the consumers which are to be tested on the electrical system of a given vehicle can be switched in a pre-defined sequence.

If the program used to control the equipment is transferred to the PLC and activated, the equipment is said to be operational and measurements can then be carried out.

2.3 Loads and Power Supply

In order to reproduce various energy suppliers, storage units and the necessary electrical consumers, different systems may be applied.

A possible strategy is the original representation of consumers by systems which are used in the vehicle. For this reason, headlight modules or window-lifter motors are integrated into the test bench and connected to the power supply and control units using the original set of cables. The necessary sensor signals and the connection to a vehicle bus-system may be realised using a rest-bus simulation. Ika already has experience in the application of necessary software tools and hardware components.

Often a replication of the physical behaviour of a load is necessary. For example, headlights can be considered as a significant ohmic load with a positive temperature-coefficient at the point when they are switched on. In case of a short circuit investigation, the PTC effect does not play a significant role, but it is sufficient to reproduce the purely ohmic part of the resistance at the point of short-circuit, using laboratory resistances. These components, whose resistances can be varied can be used to represent ohmic consumers up to a power of 3.5 kW at a voltage of 42 V.

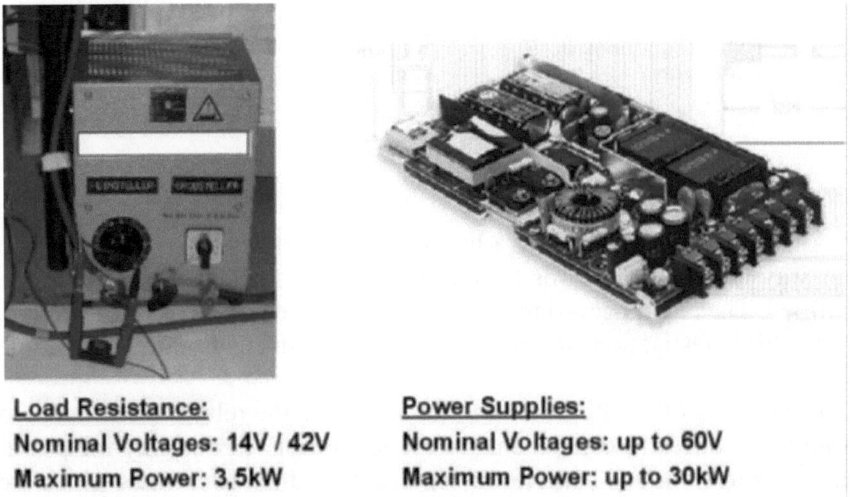

Fig. 7. Components of the test bench – loads and power supplies

In order to represent energy storage devices, batteries or supercaps are used, where in order to realise a 36 V battery, a serial connection of three 12 V batteries can be used.

Abstractions are normally used for Generators and DC/DC-Converters since either no prototypes are available or the integration of such systems in the test bench is associated with a relatively large effort. The reproduction of such components is achieved using power supplies where it is always necessary to identify the control behaviour of the voltage supply for consideration in a later evaluation.

2.4 The Climate Chamber

Apart from the setting of various parameters, which are pre-defined by the architecture of the vehicle to be investigated, relationships related to the climate have also to be often considered. Differing environmental temperatures, for example, have a large influence on the currents of batteries to be used and on the charging voltage of the generator.

In order to temper components, a specially set-up climate chamber is used. Based on requirement, heating blowers or cooling aggregates may be connected. Temperatures in the range T = -25°C to T = 100°C can hence be set.

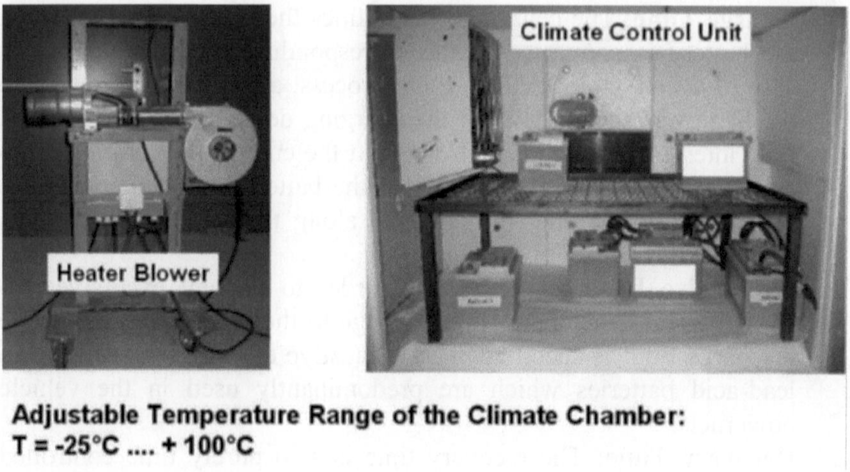

Fig. 8. The Climate Chamber

The climate chamber can be set-up in close proximity to the test bench which allows for the use of shorter battery cables. In contrast to conventional climate chambers the one used here represent a further step in the realistic representation of powernets.

2.5 The Battery Charge and Discharge Device

In order to consider the possible state-of-charge of the batteries in the investigation, a conditioning is necessary. For this purpose, an automated battery-charge-discharge station has been developed, with which, standardised and OEM-specific charge/discharge cycles can be run. The batteries are connected to the charging device for conditioning, where for different battery-parameters (capacity in Ah) respective specified connections on the charging device are present.

To charge the batteries, based on choice, conventional vehicle charging devices or controlled power supplies with current and voltage monitoring may be used. The selection of the charging device depends on conditioning requirements. In the process of conditioning, various current and voltage characteristics may be recorded, with the help of which, for example, damaged batteries may be identified.

The Software DIADEM is used in the regulation of the battery-charging-discharging station. DIADEM is used in the measurement, evaluation and visualisation of data. In order to start a charging/discharging process, the following values can for example be introduced in a DIADEM-input mask:

a. **Charge Time**: The charging time defines the time interval in which the battery is connected to the corresponding charging device. In case of a time controlled charging process, a timer is started, where the battery is separated from the charging device after the specified time interval has been run through. At the end of the charging process, the voltage across the poles of the battery is documented. The possibility of charging the batteries along the VI-characteristics is also possible.
b. **Exhaustive Discharge Limit**: In order to prevent excessive discharging and hence permanent damage to the battery, an exhaustive discharge limit is specified. The exhaustive discharge limit is 9V in lead-acid batteries which are predominantly used in the vehicle powernets.
c. **Recovery Time**: The recovery time is also purely time controlled and serves to subdue the diffusion processes and hence the recovery of the batteries. The recovery time is usually between 12 h and 24 h.
d. **Energy Extraction**: Batteries are connected to precision discharge resistances in order to discharge them. These resistances are actively cooled during the discharging process in order to prevent strong resistance changes due to heating. Using a PCMCIA-measuring card, the voltage drop across the discharging resistances is measured. Based on the voltage drop and the measured resistance values, one is able to determine the battery discharging current which is continuously integrated. Achieving the desired state-of-charge leads to the battery being separated from the charging resistance.

At the end of the charging process, the battery current is measured and documented. Finally data containing all the charging and discharging processes are documented in a battery-history. Fig. 9. shows a block diagram of the charge/discharge station.

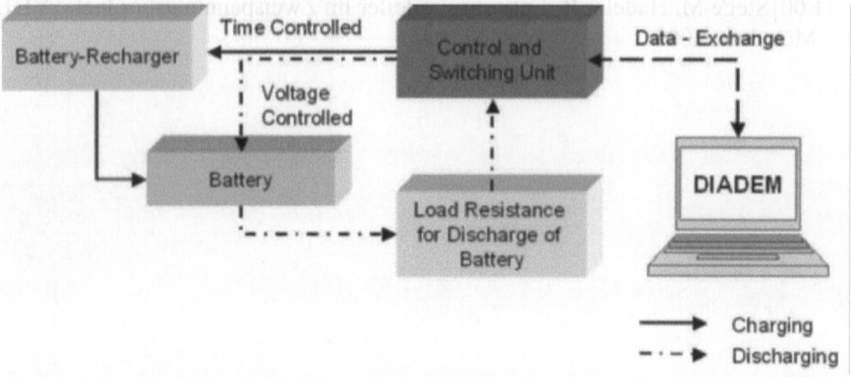

Fig. 9. Schematic of the Battery Charge/Discharge Station

Further, standardised or OEM-specific cycles are available or can be quickly realised with the help of the test bench control. A Digatron battery test bench is also available.

3 Summary

The test bench at the ika/fka distinguishes itself through the high performance of its measurement equipment as well as its high modularity. Using this test bench, a wide spectrum of real powernet architectures (energy generators/storage devices, loads, cables) can be reproduced. Using DOE, mathematical models for failures to be investigated may be determined in order to identify critical cases or to define design criteria for powernets.

4 References

[KLE01] Kleppmann W, Taschenbuch der Versuchsplanung –Produkte und Prozesse optimieren, München, 2001

[SCT00]Schöttle R, Threin G, Elektrisches Energiebordnetz: Gegenwart und Zukunft, VDI Berichte Nr. 2000, Baden-Baden, 2000

[ERI01] Erich E, Randbedingungen bei der Einführung eines neuen Bordnetzes, IIR Deutschland GmbH, Stuttgart, 2001

[GRE02] Gresch P, 42V-Bordnetz: Stand der Entwicklung, Anforderungen und Ausblick, Institut für Kraftfahrwesen Aachen, Aachen, 2002

[MOS96] Moser O, Energieverbrauch 2002 – Verbraucher im elektrischen Bordnetz, VDI Berichte Nr. 1287, Baden-Baden, 1996

[STE00] Stege M, Hadeler R, Leistungsverteiler im Zweispannungsbordnetz, VDI, München, 2000

Protection Strategies for Future Electrical Vehicle Architectures: Towards Fuseless Strategies

Joan Fontanilles, Carles Borrego, Gabriel Figuerola, Jordi Mestre

Lear Corporation, Valls (Spain)

Abstract

Today, the question is no more: «Why 42 V?» but «When?» and «How?» and the introduction of dual voltage power networks in high luxury vehicles seems immediate in a short timeframe.

New technical and economical challenges have arise, not only to the current products and technologies used in the power distribution and control systems in vehicles; but also to the whole architecture approach (high power distribution, safety-protection, EMC compliance, energy management, system monitoring, components cost, x-by-wire functions, etc).

Safety becomes critical in a system with a high voltage (up to 58 V in load-dump condition). The aim of this paper is to present new failure situations –as a result of having a dual voltage electrical network with higher voltages- in future vehicles electrical architectures. It requires defining new protection strategies to assure, at least, the same level of protection of 14 V vehicles.

1 Introduction

Power demands in automotive electrical systems have steadily increased for decades. However, within the last few years the power demand is being sharply increased. To fulfil these power requirements, it will be necessary to adopt a similar solution to the one adopted in the late 50's, when a 12 V electrical network replaced a 6 V one in vehicles. Today, a 42 V-electrical network is going to replace the 14 V one.

This transition will occur through an intermediate step to a new single power network again, with a nominal voltage of 42 V. Within this step, both current 14 V and new 42 V voltages will coexist. Obviously, evolution will increase the number of services at 42 V, which will be in the end located everywhere in the vehicle.

New failure situations –as a result of having a dual voltage electrical network- have been identified in future vehicles electrical architectures. This paper will be focused on electric arc and short circuits. For each one, protections proposals will be presented.

When the integration in a vehicle will be completed and fully tested, pros&cons will be pointed out.

It is worth mentioning there is a common effort among carmakers and suppliers to investigate together these specific problem: Within the forum Vehicle Electrical Systems it was decided to establish a Working Party Short Circuits (WPSC) to handle the short circuits problem. Parts of the results of this work are reflected in this paper. Within MIT Consortium, there is a RU#7A has the objective of investigating differences in electric arcs at 42 V and at 14 V including periodic recurring arcs and opening of connectors under load.

2 Electrical architectures

Experience shows there is not an universal solution for protection strategies, valid for any electrical architecture, and each solution relies on the own characteristics of power distribution and control. It is difficult for current single voltage architectures, now, for dual voltage architectures protection becomes a challenge due to possible failure modes are increased (Fig. 1.).

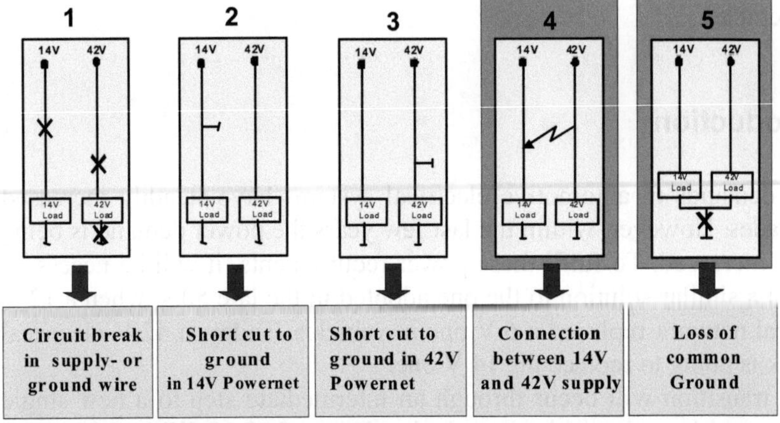

Fig. 1. Failure modes in a Dual Voltage Architecture.

2.1 Today's Circuits Protections

A short circuit of a power line to ground (cases 2 & 3) can be handled as today. However, conventional circuit protection methodology (e.g. circuit breaker, fuse) does not assure the protection against the new failure modes, (cases 4 & 5).

Today's protection devices are not capable of detecting and protecting short circuits between 14 V and 42 V busses. So, the affected wires and terminals might be overloaded and damaged. Depending on the architecture, the 12 V Battery could have a negative over voltage. This over voltage might destroy the 12 V Battery.

In a Loss of Common Ground situation, it appears a reverse current downstream the 14 V loads and a reverse voltage of around 28 V on the 14 V bus. The protection device on the 42 V side may not protect, and the fuse on the 14 V side may not blow. So, the electrical system is not protected.

2.2 Proposed New Architecture for Circuits Protection

Lear developed and tested in real conditions the first in-vehicle dual voltage architecture concept in 1999 [2]. It highlighted first benefits and inconveniences of the coexistence of two main voltages in one electrical and electronic distribution system. It allowed determining the first design guidelines for a new improved concept, which considers the overall electrical and electronic 2005-2010 vehicles requirements, and the 42V-PowerNet from the earlier conception.

This Dual Voltage architecture proposal, called CAVA [3], (Common Advanced Vehicle Architecture (Fig. 2.), implements several new features, as Dual Voltage Smart Distribution Nodes (DV-SDN) with embedded DC/DC converters, battery monitoring features, mechatronics, hierarchical busses communication, etc; and also introduces new circuit protections accordingly:

- Prevention of 42 V-14 V short-circuits. It can be done by reinforcing physical separation of 42 V and 14 V power circuits (Mechatronics) and mechanical protection (taping, channels, tubes). It must ensure direct short circuit from 42 V to ground instead of 14 V short circuit.
- Detection & protection of 42 V-14 V short-circuits. Specific intelligent devices used for the battery monitoring and DV-SDNs share their smart semiconductors and sensors. It is important to highlight that electromechanical relays are not recommended for these purposes because of higher disconnection time.

- Loss of common ground. 14 V and 42 V ground circuits shall not be connected to the same physical ground location. Grounds are separated in a single voltage vehicle to avoid noise and transient issues from mixing power and electronic. Now, it is necessary do something similar to avoid loss of common ground. Notice disassembly operation are probably the most critical.
- 42 V arcing. Specific intelligent devices used like DV-SDNs can share their smart semiconductors to prevent any damage during a hot disconnection. It is important to highlight that electromechanical relays are not preferable for this use due to the higher time of reaction.

The protection strategies against short-circuits and arcing presented along this paper, are based in this new electrical architecture concept.

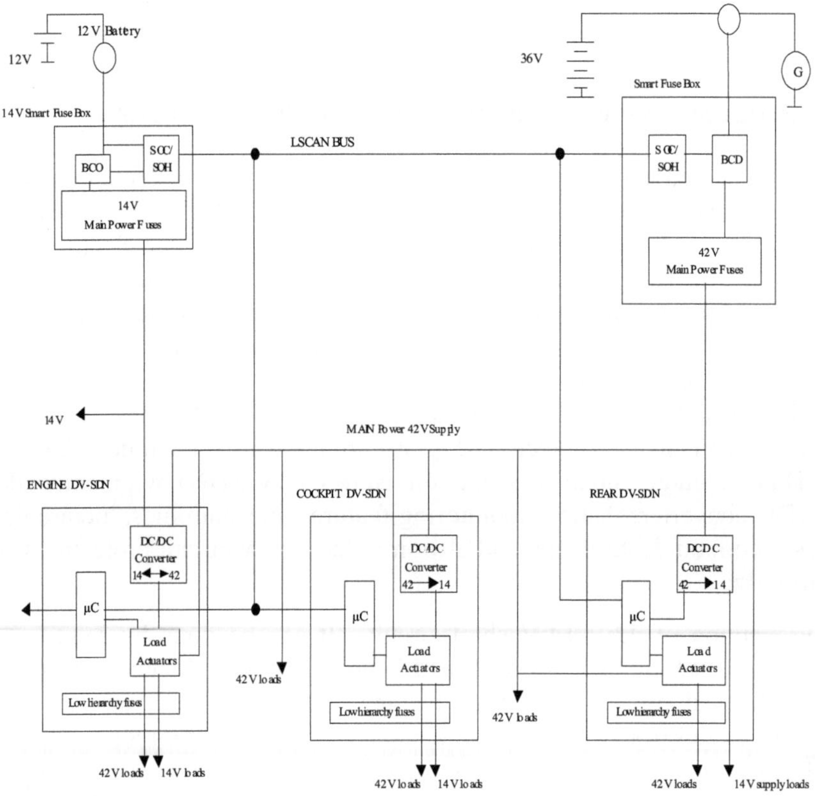

Fig. 2. Basic CAVA Power Distribution

3 Short-circuits

Short-circuits affects to the safety of the electrical system in a vehicle. In coming dual voltage architectures are identified two types of short circuits: positive to ground, the traditional one, and between positives (+14 V and +42 V), the new one. This second one is a direct consequence of having two voltages in the system.

Fuses protects any short-circuit to ground. However, traditional conceptions for system protection can not fully grant the safety in dual voltage systems, where consequences can be even more hazardous. In following paragraphs it is described a possible solution.

To realise about the scope of this problem, only mentioning the recommendation from Sci-Worx of "avoiding the system becomes unstable" or the specific workgroup about this subject within the same forum.

3.1 General aspects

Short-circuits must be prevented. Consequences must be minimised. Unfortunately, there is not an universal solution. A solution relies on the characteristics of power distribution and control of each electrical architecture concept.

The new architecture concept for 2005-2010 vehicles developed at Lear has been the reference scenario to design and implement suitable protections proposals to make a "short-circuits-proof architecture". Particular characteristics of this architecture like "power zoning" and "conversion zoning", distributed control and Smart Distribution Nodes are essential for the success of the proposed countermeasures, reducing short-circuits occurrence and allowing detection and facilitating a fast disconnection. The system is completed offering several configurations for re-connection of non-affected loads.

Simulation is a very helpful tool for these purposes [4].

3.2 Solutions for short-circuits between positive poles

Several particular areasspecific areas (Fig. 3.) can be identified in a vehicle, which the own failure characteristics, and then, the suitable protection can be implemented. Any direct connection path between batteries have to be avoided by using circuit breakers where and when necessary.

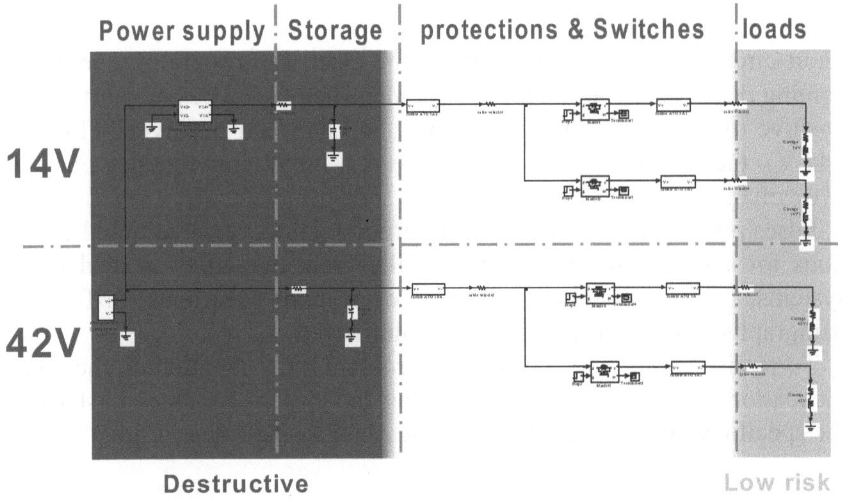

Fig. 3. Short-circuits areas

1. Short-circuits between SDNs and power generators (36 V battery and Integrated Starter Alternator or belt-driven Alternators). Mechanical protection isolating 42 V and 14 V lines is advisable. Protection is only guaranteed after battery disconnection.
2. Short-circuits between SDN and loads. The short-circuit is always established after the SDN. Protection is only electronically guaranteed. This situation is taken as the basis for short-circuit protection described in this point. Nevertheless, for a complete protection any bus physical layer and any electronic module input must be fool proof in front of a short-circuit of 42 V-14 V. It will be achieved modifying current designs at component level.

Two types of short-circuits are identified according to the resistance of the short-circuit current path between the two networks: **high impedance** and **low impedance.**

In a **low impedance short-circuit**, the resistance of the involved wires is usually between 50 mOhm and 300 mOhm, so currents may vary between 480 A and 80 A (or lower, depending on different elements in parallel and the status of charge of the battery).

This overcurrent may blow some fuses (removing the short-circuit), burn wires or the 12 V battery could even explode. Depending on the combination of blown fuses, 42 V loads can be then permanently connected to a 14 V load with supply from 14 V network (permitted) or vice versa (forbidden). A steady overvoltage applied to sensitive 14 V elements may damage these elements, so a fast disconnection is needed. Anyhow, some 14 V elements, like busses should be prepared to withstand higher overvoltages than today.

The classical protection elements like fuses may not totally interrupt the short-circuit path, so they are inefficient.

In a **high impedance short-circuit** the amount of energy between networks is like a leakage current. Values may vary between 10 to 30 A. The overvoltage produced in the 14 V network doesn't damage sensitive 14 V elements due to there is a big voltage drop across the short-circuit path. The classical protection elements like fuses may not blown due to the low currents involved, so they are inefficient.

Definitely, a complete new protection system is needed in dual voltage architectures to protect wireharnesses of all pernicious effects of short-circuits between networks.

The protection features described below rely on solid-state smart switches especially in the 42V-PowerNet.

The steps for a full protection are: **Detection** of a short-circuit situation, **Identification** of the 42 V load involved in the short-circuit and **Disconnection** of this load from the system, removing the short-circuit situation.

Detection (time is around 1.5 ms) sensing voltages and currents at different points within both power networks and a fast processing considering pre-fixed thresholds (abnormal conditions) short-circuits can be detected with enough time to react. As soon as a short-circuit is detected, it starts the **Identification** procedure of the 42 V load involved (the SDN receives an interruption in 300 µs and it identifies the involved load in between 500 µC and 5 ms). Then, the **Disconnection** should be shorter than 1 ms. So, the latency of the final solution is between 3 ms and 8 ms. These values are highly dependant of the technology used for the µC, the solid-state switch and the physical elements involved.

The protection method described requires information from the system like DC/DC converter status, current drained and supplied to/from the batteries and voltage in both networks. From these information is quite simple to identify a low impedance short-circuit, and then it is applied a simple algorithm of disconnection and re-connection (depending on each OEM strategy). With this strategy it is detected the load or loads involved in a short-circuit, disconnecting them and finally reporting the fault to the diagnostic system.

When identified the load, it is possible that it works supplied from 14 V. However in most of the cases, the 14 V fuse could blown or, in case of controlled by a solid-state switch, it can be disconnected via the 14 V switch. At the end the worst problem is a 42 V load continuously working while the 14 V is connected.

High impedance short-circuits are quite difficult to detect because they are more an overload for the 42 V load controller rather than a short-circuit. For this reason, it is important to monitor the current of the load to know when the load is working under abnormal conditions.

A possible solution is to implement, using a μC, a behaviour pattern for the load considering inrush currents, normal operation conditions, variations of the input voltage, speed, etc. If a certain amount of current is exceed this could be identified as a leakage current from the 42 V to the 14 V powernet. Anyhow, this is not easy for example for new complex high consumers that are expected to be in a vehicle or if small currents wanted to be identified.

A Battery Disconnection Switch (BDS) would provide the highest level of protection to the system when a short-circuits occur. Any short-circuit path is cut from the origin. In specific situations (dealer, manufacturing line,...) the complete system can be totally disconnected to avoid any damage.

Examples of short-circuits situations with different protection strategies implemented are described below

Fig. 4. depicts a short-circuit without protection. The [2] plot represents the 42V-PowerNet (in AC, with 2.5 V/div scale) and the [1] line the 14 V powernet (in AC, with 3.5 V/div scale).

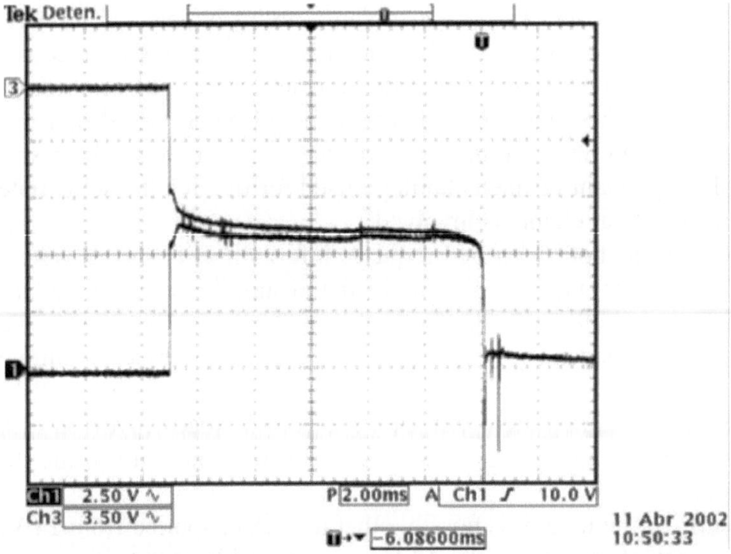

Fig. 4. Example of a short-circuit between powernets protected by fuses

This situation is a shortcircuit of a 42 V cooling fan protected by a 15 A fuse (ATO size) with a 14 V heating windshield protected by a 30 A fuse (ATO size). Both loads are supplied by a 6 mm^2 wire and the shortcircuit load is established at the connector level by another 6 mm^2 wire. The 15 A fuse blows in 11 ms due its low I^2*t in front of the 14 V line fuse. This time highly depends on the battery status (state of charge) and on the ohmic resistance of the shortcircuit.

In this case, the cooling fan will be permanently connected to the heating windshield without blowing the 30 A fuse. It would be disconnected only switching off the heating windshield. The 14 V powernet will be submitted to an overvoltage of 8 V (depending on the test point) during 11 ms, so, it could damage some electronic components.

It is obvious that a faster and smarter protection is needed. Fig. 5. depicts a protection system based on the voltage processing of both networks.

Now, the shortcircuit is interrupted in 1.34 ms (ten times faster than using a conventional fuse protection). The system identifies also the load involved, disconnecting it from the system. Notice solid state switches control both circuits now, there are no fuses and the abrupt waveform and the disconnection of the system is due the inductance effects of the wires in front of high variations of current.

Using other information available in the system, especially from batteries and from the status of the converter, the shortcircuit protection can be improved, as seen in Fig. 6. The disconnection time is near 660 μs, even faster than in the previous cases.

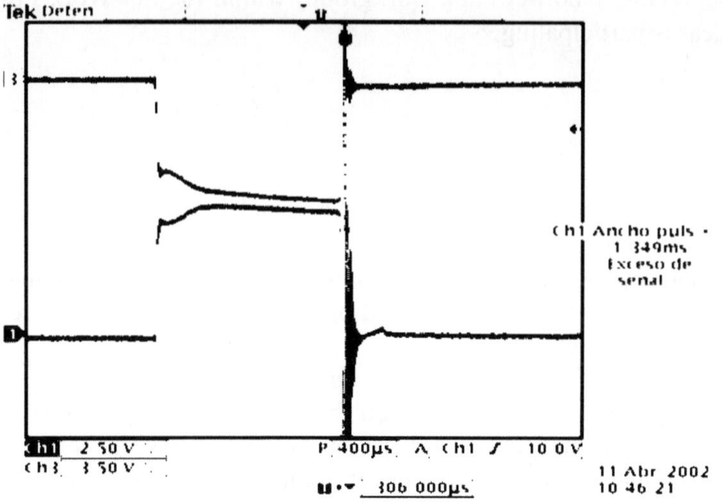

Fig. 5. Example of a smart short-circuit protection between powernets

In high impedance shortcircuits, the process of the load pattern increases the detection time depending on the load (a windshield is easier modelled than a cooling fan or a power window) but the philosophy of protection to implement is the same.

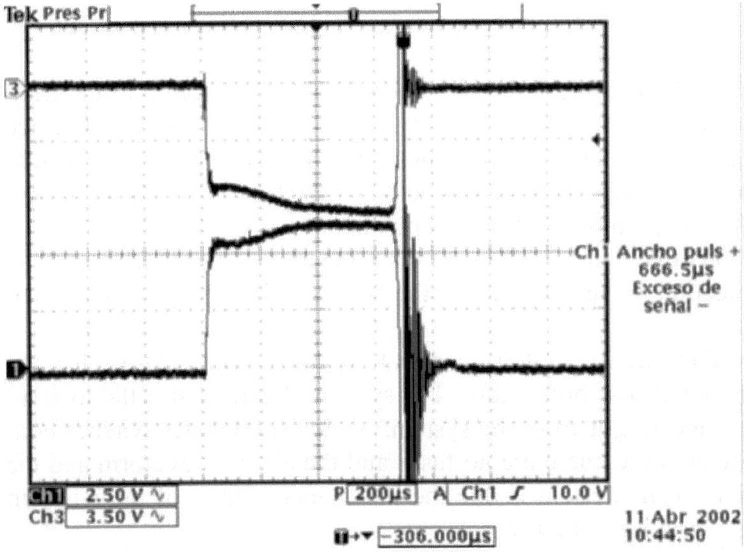

Fig. 6. Example of a short-circuit between powernets protected by smart systems using as information from the SDN and the system.

This work, still in progress, is being done under the umbrella of the activities of the WGSC (short-circuits workgroup) within the Sci-Worx Forum where Lear is participating.

4 Arcing

Thousands of circuit interruptions occur in the vehicle's life like stopping the vehicle, switching off a service or even seldom blowing a fuse or changing the battery. A connector, an electromechanical or a solid-state switch or a protection device (fuses, circuit breakers, etc) are the direct responsible to open the electrical circuit.

They are designed to interrupt the current path opening a metal-to-metal contact. From physics, an arc fault -current bridging an air gap- is generated during this circuit interruption till the arc voltage across the contacts becomes greater than the source voltage. It could happen at any voltage, but fortunately, in today's vehicle, in most of the cases the arc is not noticeable because its duration is very short (less than nanoseconds). However in higher voltages (>20 VDC), even currents can be smaller, arc becomes much more long and severe.

Arc effects are worse in inductive circuits than in resistive ones. This is due to the energy stored in the inductive circuit that helps to maintain the arc till the total energy dissipation.

Two types of arcs are identified:

1. Serial arcing: It occurs on disconnection of connectors, loose lugs, terminals or broken wire. This fault is in series with the load.
2. Parallel arcing. It occurs when damaged wires touch the chassis (ground) or between two adjacent cavities in a connector. This fault is in parallel with the load. Parallel arcing can be easily associated to short-circuit. Following paragraphs within this point are devoted to serial arcing.

Arcing effects are pernicious and hazardous: components, wires and contacts can be irreversible damaged. Minimise risk of hot unplugs is a must, and traditional protection devices are not effective.

4.1 Solutions for serial arcing prevention

Connectors are plastic cavities grouped. In each one the terminals assure the electrical continuity of an electric circuit.

During a disconnection under load, the arc can appear in several pairs of male-female terminals of connectors that can be damaged.

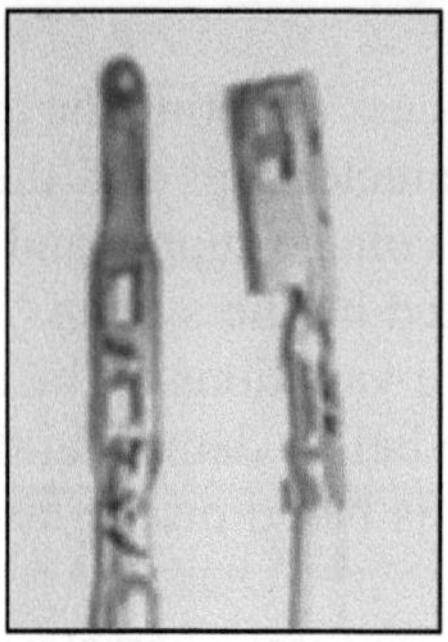

	50mm/Min	100mm/MIN	500mm/MIN
15A	damage	damage	damage
10A	OK	OK	OK
5A	OK	OK	OK

Fig. 7. Arc Voltage vs. Arc Current

An arcing is determined by a threshold voltage (depending on the material of the cathode) and a threshold current (depending on the material of the anode). Under the thresholds the arcing is not stable (Fig. 7.).

A SDN's implements a feature that senses the "connector unplugging intention" (Fig. 8.) to prevent arcing from any disconnection of loads controlled from it. These signals are detected and processed by the µC in the SDN locking the actuators of the loads driven from the affected connector. It permits a safety unplugging of the connector with the circuit previously switched-off. High energy arcing is avoided.

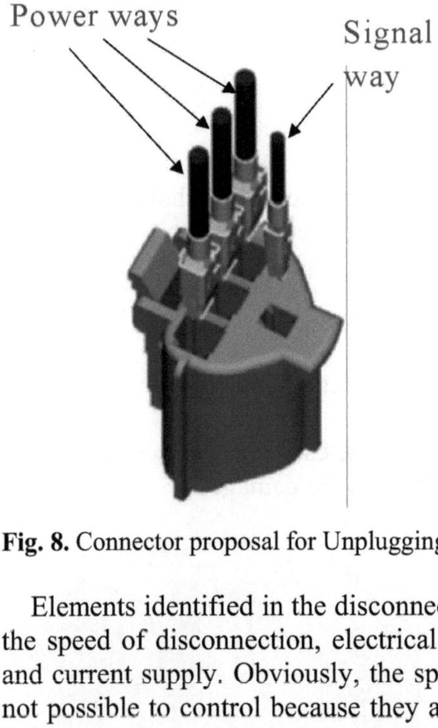

Fig. 8. Connector proposal for Unplugging disconnection detection

Elements identified in the disconnection process affecting the arcing are the speed of disconnection, electrical load characteristics and the voltage and current supply. Obviously, the speed, the voltage supply and load are not possible to control because they are prefixed by the system. So, using indirect measures of the current and the voltage variations in the terminals it has to be possible to prevent a disconnection before damaging the terminals or the connectors. The protection method is based on the fast disconnection of the load power supply.

Following scopes depicts a terminals and connector characterisation in disconnection with load at different voltages and disconnection speeds.

In Fig. 9. and Fig. 10., top trace corresponds to the 42 V supply voltage (DC scale 10V/div) and bottom trace corresponds to current (AC scale 10 A/div left and 5 A/div right). The positive slope of bottom trace corresponds to the disconnection process for a 10 A load left and 5 A load right. The noise in the current measurement is due the sensor technology.

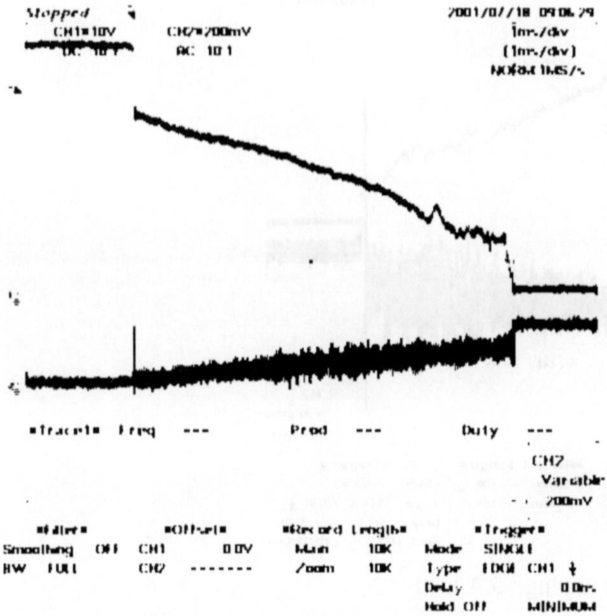

Fig. 9. Example of hot unplug: 10 A load

In all the cases there is a voltage drop of near 10 V. It is supposed that this voltage is due the ionisation of the air due to the presence of a potential afterwards the voltage and the current decrease with a defined slope. The duration of the slope, in direct relationship with the inductance and the type of load, determines the duration of the arcing. In previous cases this arcing has duration of 6.5 ms and 3 ms. Energies involved during the arcing have values higher than 2 Joules depending on the voltage and the current.

The previous waveforms have the necessary information to identify an unplugging in front other situations that can appear in a car (voltage variation, load changing) and could probably mask this effect. Lear has determined two approaches to identify this arcing situation and may disconnect the circuit in μs if all conditions are established.

Fig. 11. and Fig. 12. represent the disconnection process using Lear's methods to identify an arcing. Top trace corresponds to the voltage in the terminals and bottom trace corresponds to the signal that disconnects the FET that controls the load. The left figure corresponds to a resistive load of 15 A. The right one to an inductive load of 15 A.

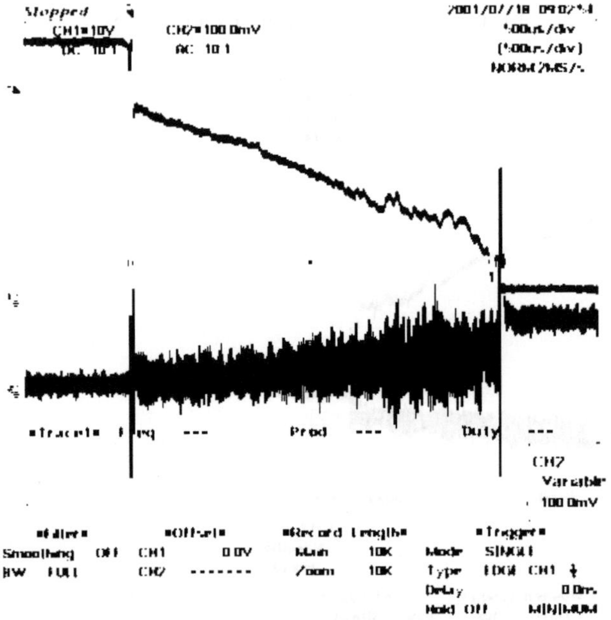

Fig. 10. Examples of hot unplug.: 5 A load

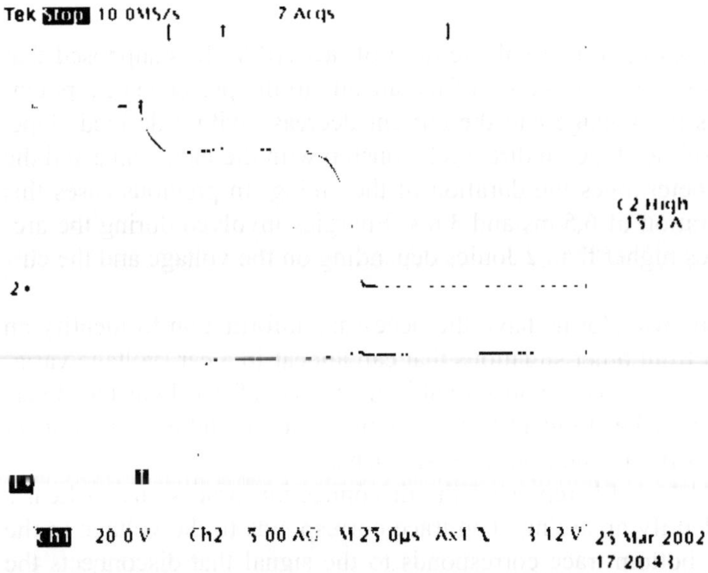

Fig. 11. Example of hot unplug: 15 A resistive load

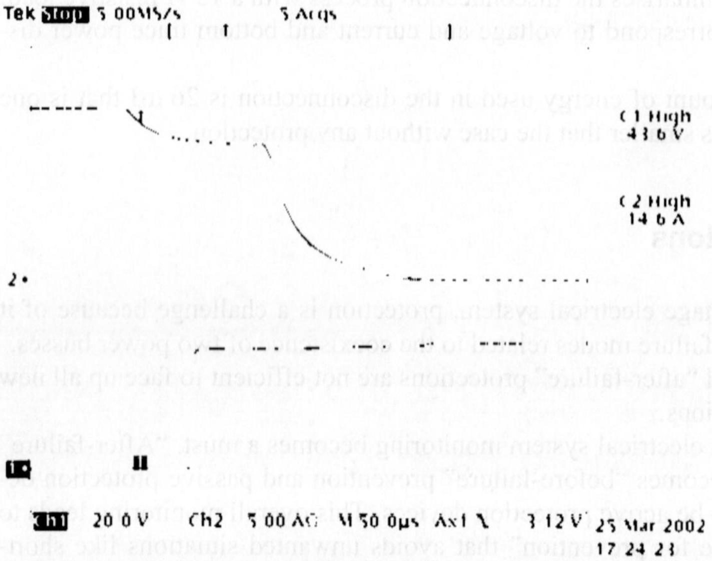

Fig. 12. Example of hot unplug: 15 A inductive load

The detection of a disconnection is done in less than 25 µs and the disconnection of the FET before the arc is set is done in 100 µs, so this small time cannot damage the terminal. In the right figure this time is increased due to the inductive characteristic of the load and the time is near 250 µs.

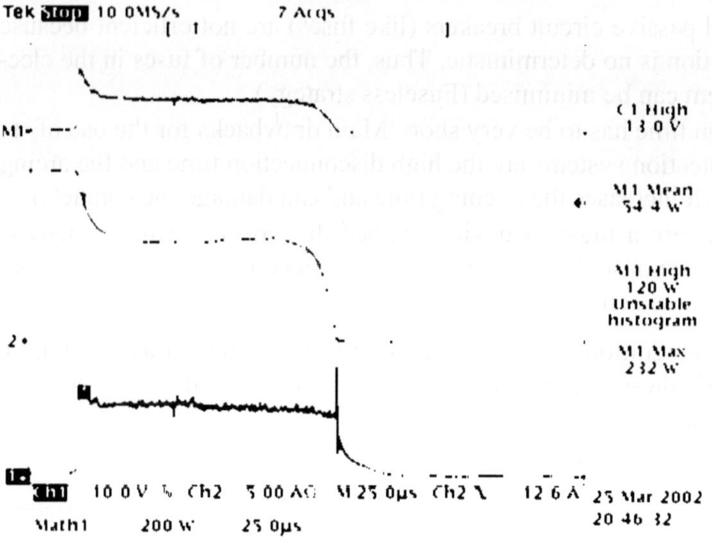

Fig. 13. Power involved in the disconnection

Fig 13. summarises the disconnection process with a 15 A resistive load. Top traces correspond to voltage and current and bottom trace power dissipation.

So the amount of energy used in the disconnection is 26 mJ that is one hundred times smaller that the case without any protection.

5 Conclusions

In a dual voltage electrical system, protection is a challenge because of it appears new failure modes related to the coexistence of two power busses.

Traditional "after-failure" protections are not efficient to face up all new critical situations.

An overall electrical system monitoring becomes a must. "After-failure" protection becomes "before-failure" prevention and passive protection devices need to be active protection devices. This overall monitoring leads to a "knowledge for prevention" that avoids unwanted situations like short-circuits or arcing damage.

Experience shows there is not a universal solution for protection strategies, valid for any electrical architecture, so it has been presented a proposal relying on the own characteristics of power distribution and control. When the integration in a vehicle will be finished the validation will be completed.

Main highlights from this research are:

- Traditional passive circuit breakers (like fuses) are not efficient because their actuation is no deterministic. Thus, the number of fuses in the electrical system can be minimised (Fuseless strategy).
- Interruption time has to be very short. Main drawbacks for the use of relays in protection systems are the high disconnection time and the arcing in contact (it increases the opening time and can damage the contacts).
- Diagnostics are a must. If devices embed diagnostics features increase the system response, because any abnormal operation is reported immediately to the system.

When the integration in a vehicle will be finished the validation will be completed. All these features must be worked out from the early conception of the architecture.

6 Definitions, Acronyms, Abbreviations

SDN:	Smart Distribution Node
SFB:	Smart Fuse Box
µC:	Microcontroller
µs:	Microseconds
ms:	millisecond
mOhm:	milliohm

7 References

[1] WORKING PARTY SHORT-CIRCUIT DOCUMENT, System Protection Guideline for Dual 42V/14V Vehicle Electrical System, Working document, 2002

[2] Borrego C, Fontanilles J, Giró J; (Lear), Dual-Voltage EEDS Strategy, 42V Powernet, Villach (Sep. 99)

[3] Borrego C, Figuerola G, Fontanilles J, Giró J, Mestre J; (Lear); 42 V Power Distribution Network For Future Vehicle Generations; Chapter of the Book: THE NEW AUTOMOTIVE 42V POWERNET; ExpertVerlag 2002.

[4] Gohin G (PSA), Meurant C (PSA); Fontanilles J; Borrego C (Lear), Dual Voltage Architecture Optimisation Through A Virtual Simulation Platform SAE ATT Barcelona (2001), SAE 01ATT25101

[5] Graf A, Estl H (Infineon), Fuse replacement with Smart Power Semiconductors, 8th International Conference Vehicle Electronics, den Baden 1998

42V PowerNet System Protection Concepts

Edmund Erich

Delphi, Wuppertal

Abstract

Introduction of the 42V-PowerNet requires consideration of architecture and voltage specific fault scenarios.

It is the intention of this paper to create awareness of the most critical issues (corrosion, arcing and inter-bus short circuits) and to show solution concepts depending on application and the chosen architecture.

The trend towards semiconductor based power switching will support adaptive system protection and diagnostics for monitoring the state of the electrical system.

When we discuss the planned implementation of higher system voltage into future vehicles we always meet the topic of safety.

First of all there is the issue of human protection. This has been solved by choosing a voltage level which is considered safe for human beings. Therefore additional safety measures are not really required. Vehicle safety is the second issue:

- What is the impact of 42 V on the vehicle elctrical system?
- If we apply a dual voltage architecture - will the 14 V system be safe?
- How do chemical and physical effects like corrosion and arcing behave at 42 V?
- What are the recommended measures to be taken?

The following paper will give some answers to these questions.

1 Introduction

After the euphorical start of the discussion on the new “42 Voltage” level, we are now developing the next generation of products to guaranty the realization on the same quality expectation for future automobiles. The introduction of a higher voltage level requires additional technical solutions as well as higher costs without any perceptible benefit/value for the consumer.

On the other hand we are confronted with small but challenging details which, for the time being, make an engineer's life exciting.

When introducing a "42 Voltage" System, we assume that the vehicles have two voltage levels, i.e. a 14/42 "Dual Voltage E/E Distribution System". Therefore, we would like to examine the following topics more closely:

1. Corrosion within the E/E system and especially within the related connection systems,
2. Emergence of sustained electric arcs and
3. Protection concepts for these two voltage busses against each other.

	>42 Volt not distributed	42 Volt	12 Volt	stab. 5 Volt not distributed
today			all Loads	Electronics ⇨ local Voltage Controllers
Level A	i.e. Electric Valve Train ⇨ local DC-DC Converters	new Systems i.e. E-Cat and high power Loads like Starter, Pumps, Ignition	all remaining Loads	Electronics ⇨ local Voltage Controllers
Level B	⇨ local DC-DC Converters	all remaining Loads	Lamps	Electronics ⇨ local Voltage Controllers (DC/DC Converter at 42V)
Level C	⇨ local DC-DC Converters	all Loads		Electronics ⇨ local DC-DC Converters

Fig. 1. Migration-Scenario from 14 V-Electrical System to 42 V

2 42V-PowerNet: System Protection Concepts

The general classification of dual voltage architectures is based on some following criteria:

- Technology of generator
  Output voltage stage(s)
- Amount of batteries
  Monitoring and Energy management
- Redundancy of energy supplies

Safety critical functions

- Availability of 42 V components (DC/DC; actuators & sensors)
- Cost of additional needed products and technoologies

The following examples represent two of the mainly considered dual voltage energy supply architectures.

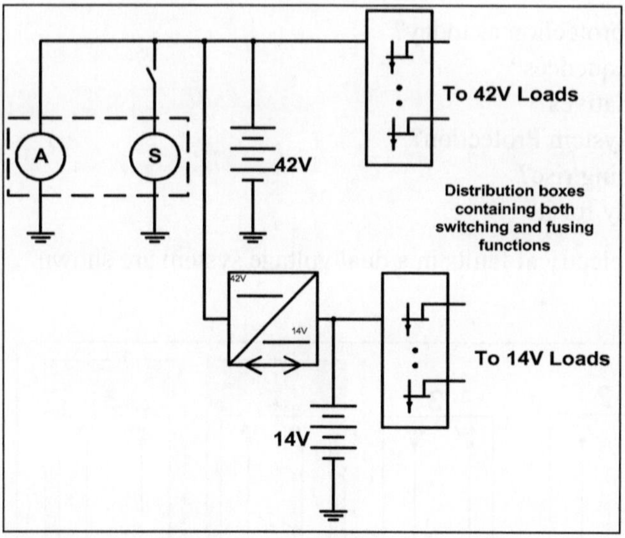

Fig. 2. One Generator - two Battery System with central DC/DC-Converter

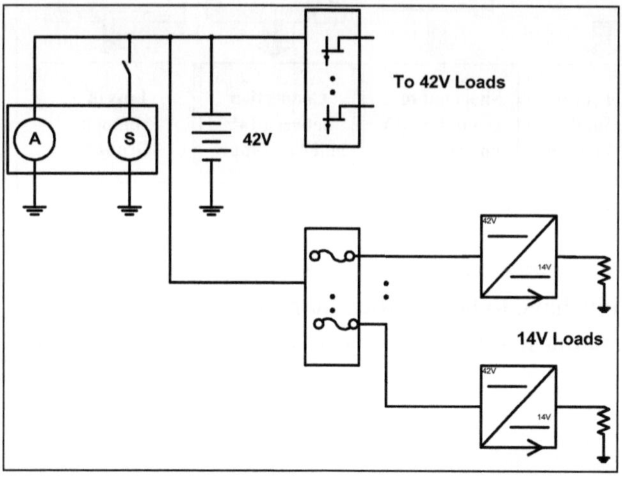

Fig. 3. One Generator/one Battery System with decentral DC/DC – Converters

3 General Questions: What degree of System Protection is required? What is the cost margin?

Compared to today's vehicle electrical system we have to face a new situation when implementing the 42V-PowerNet. Due to the higher voltage and the special characteristics of a dual voltage system some of the effects are more aggressive than today. The arising questions are:

- Can we do circuit protection as today?
- What are the consequences?
- What are the alternatives?
- Why do we need System Protection?
- What is the remaining risc?
- Who is going to pay for it?

In Fig. 4. relevant electrical faults in a dual voltage system are shown.

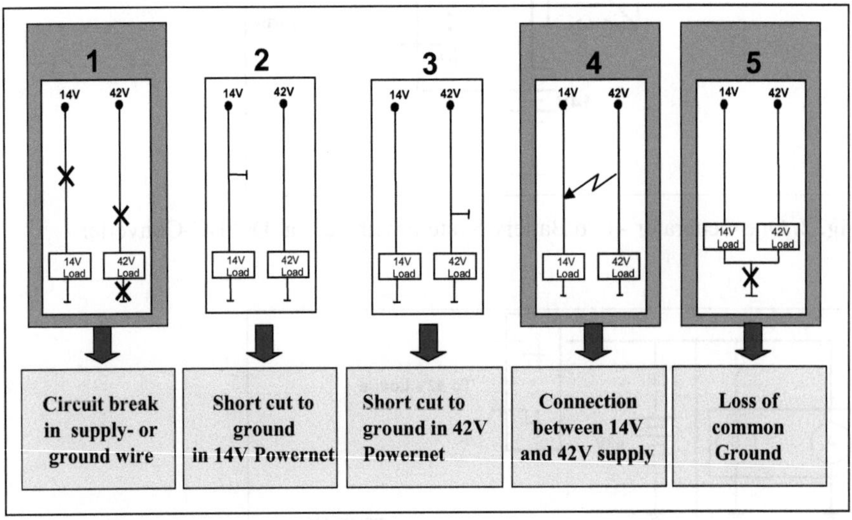

Fig. 4. Electric faults

Especially three faults have to be regarded more intensively.

In order to avoid faults a proposal for a dual voltage architecture is shown below.

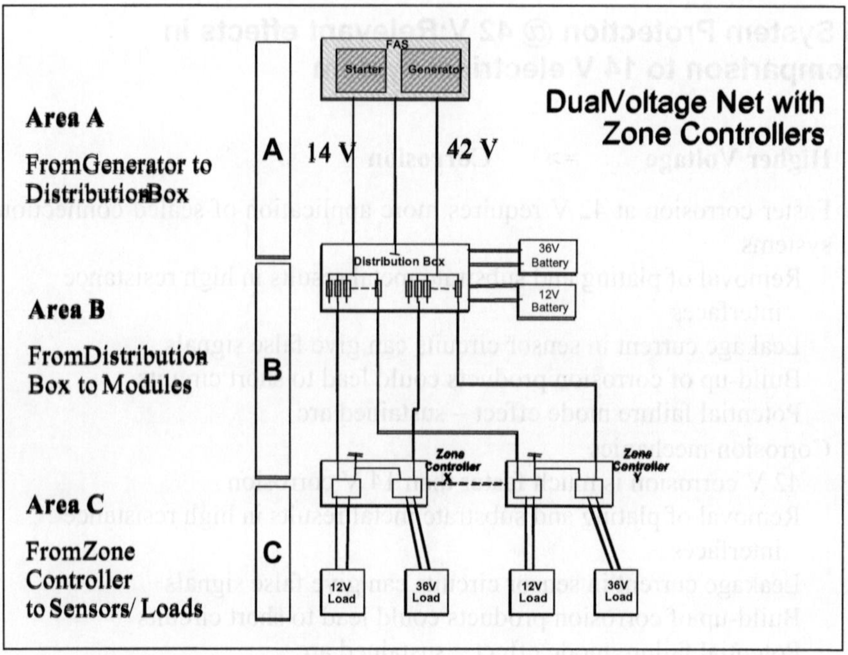

Fig. 5. Vehical Electrical System

4 Effects in 42V-PowerNet / Corrosion

Due to the increased electrical field intensity there is a higher possibility of corrosive processes to arise between the contacts under 42 V conditions in a humid environment. Furthermore, the corrosion process itself generates metal oxides which can lead to conductance between the terminals and finally cause possible short circuits. This phenomenon has been proved in several tests and has been documented by a video. In order to avoid these impacts, the following steps are recommended:

- Cavity isolation through silicon sealing,
- Inside the connection system,
- Outside the connection system (sealing to the wiring harness),
- Extension of the creepage distance between the terminals,
- Increase the cavity distance and
- Other constructive steps.

In general we recommend to seal all 42 V connectors as stated above, no matter if they are used in a humid environment or not.

5 System Protection @ 42 V:Relevant effects in comparison to 14 V electrical system

Higher Voltage => Corrosion

- Faster corrosion at 42 V requires more application of sealed connection systems
  - Removal of plating and substrate metal results in high resistance interfaces
  - Leakage current in sensor circuits can give false signals
  - Build-up of corrosion products could lead to short circuits
  - Potential failure mode effect – sustained arc
- Corrosion mechanics
  - 42 V corrosion is much faster than 14 V corrosion
  - Removal of plating and substrate metal results in high resistance interfaces
  - Leakage current in sensor circuits can give false signals
  - Build-up of corrosion products could lead to short circuits
  - Potential failure mode effect – sustained arc

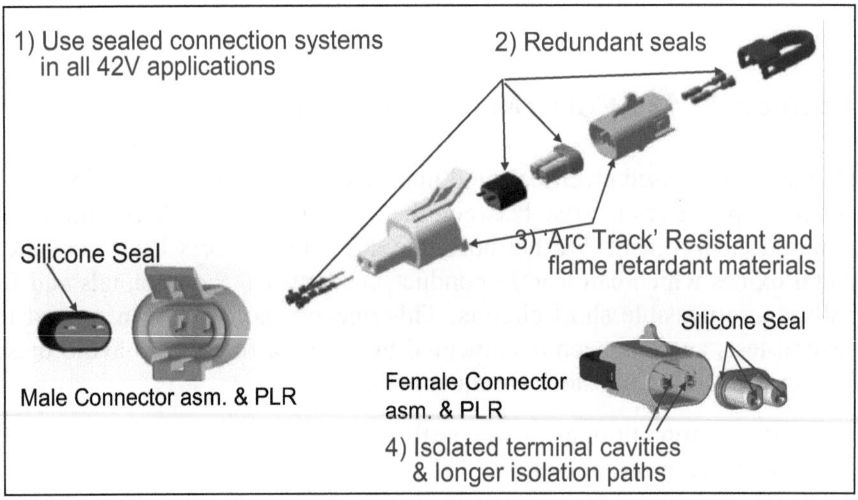

Fig. 6. Key design features for corrosion protection

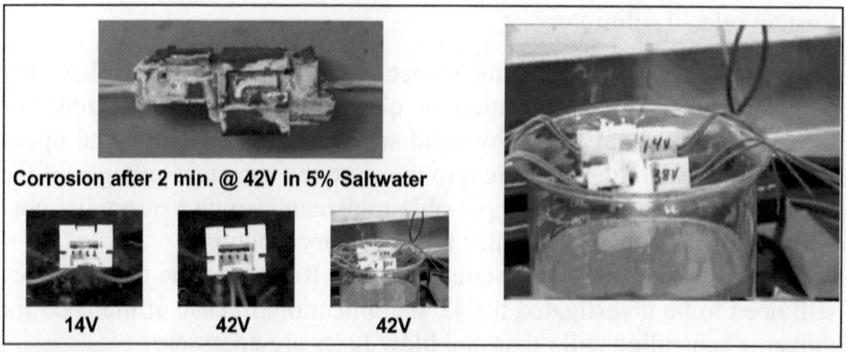

Fig. 7. Effects in 42V-PowerNet / Corrision

6 Effects in 42V PowerNet / Arcing

The emergence of sustained electric arcs basically depends on the voltage, and does not create any problem in a 12 V network under normal circumstances. Voltages above 18 V tend to have high energy electric arcs which may damage the metal contacts or even destroy them. We recommend a limit of 10 joule arcing energy max. If the electric arc's energy is below 10 joule, it seems to be in an acceptable range. Still necessary validations will show how the OEMs' quality requirements can be met.

7 Connection system related solutions

Connection system related solutions are:

- forerunning terminals to predict an upcoming unplugging
- In case of unplugging the terminals for diagnosis will be opened first. The central electronic detects the action and the load current will be switched off before the power terminals will be unplugged. This means additional content at the connection system, at the central electronic as well as at the wiring harness
- Constructive control of the disconnection speed to reduce the arcing energy
- Introduction of new contact materials
- Use of gasing materials
- Elongation of the arc by use of magnetic fields to reduce the partial energy

System related solutions:

- Future vehicle E/E architecures (especially 42 V systems) will have the tendency to a higher utilization of electronics. Relais and traditional blow fuses will be replaced by solid state technology. This trend opens the possibility to the E/E system to measure the currents/ voltages of the related loads and in addition possible transients, which can predict possible failors and finally initialize the necessary reaction.
  Current sensitive FET´s are state of the art. Reaction time and accuracy still need to be investigated for 42 V application. In case of the need for galvanic separation still relais and blow fuses are necessary.

In any case each load circuit should be switched and fused based on its individual requirements. In some cases both, the system protection and at the same time the component related solution are necessary to guaranty safe conditions and operation.

8 System Protection @ 42V

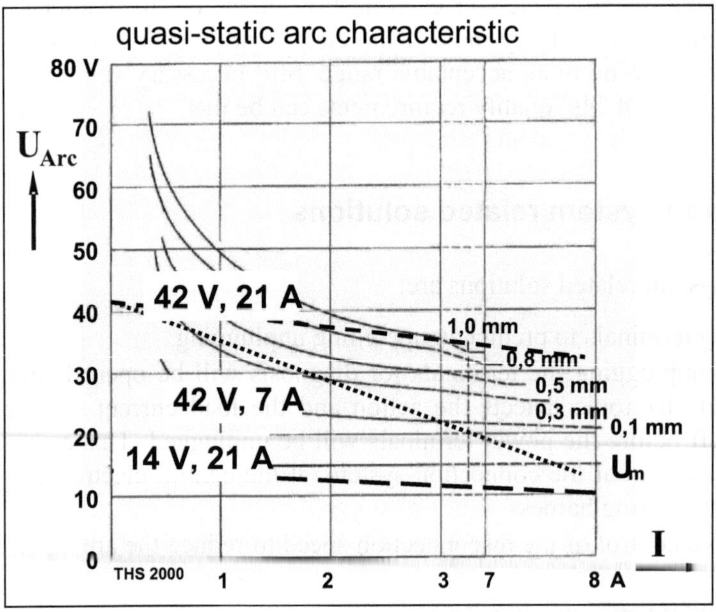

Fig. 8. Effects in 42V-PoweNet/Arcing

Relevant effects in comparison to 14 V electrical system

Higher Voltage => Arcing

- At voltages higher than 20 VDC arcs might not be self-exstinguishing. For instance at 42 V and an inductive load current of 10 A the arc may stay alive up to a gap of 3 mm.
- Arc extinguishing lengths are much longer with 42 V

Serial arc

- broken wire
- loose connector
- mating & unmating of connectors with active load (hot plugging)

} resulting current lower than load current

Parallel arc

- intermittent short to ground

} resulting current depending on various parameters and difficult to predict

With the exception of low energy hot plugging, system solutions for arc detection are required.

Fig. 9. Effects in 42V-PoweNet/Arcing

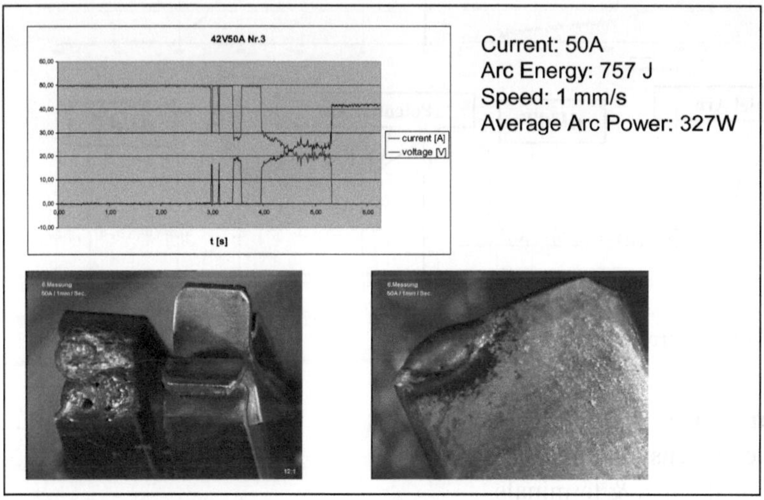

Fig. 10. Hot Plug/Unplug

- Hot Plug/Unplug
  Opening a connection under load will generate an arc
  Energy is a function voltage, current and time
  Arc energy at 42 V might be sufficient to damage terminals

9 Arcs @ 42V / Hot Plugging

Potential solutions for design - and material improvement

Component level

- Connector with controlled speed
  Low additional cost
- Gasing materials
  Low additional cost
  (Cents per wire on basis of sealed connection systems)
- Magnetic arc extinction
  High cost and low currents

Potential solutions for design- and material improvement

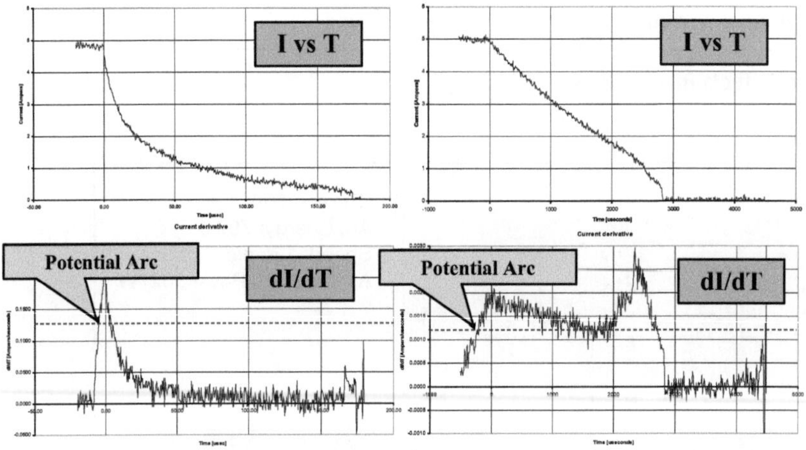

Fig. 11. Arcing current signature

System Level

- Additional sensing contacts
- Additional wires & terminals
  => Relatively low additional cost

- Arc Detection
  Detection Electronics
  Wire fault protection
  => Relatively high additional cost

Based on future E/E electronics for fusing we expect additional detection of voltage and current transients.

- Arcing current signature detection

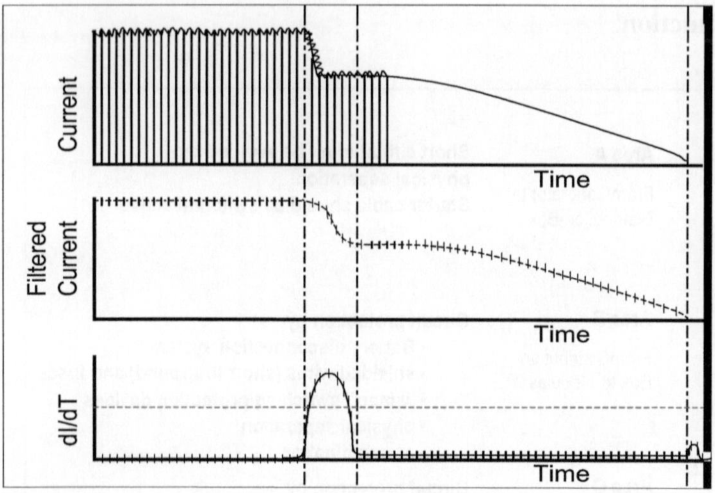

Fig. 12. Arcing detection

10 Effects in 42V-PowerNet / Short Circuit

- System protection @ 42 V
  Relevant effects in comparison to 14 V electrical system

  Dual voltage system => **14 V/42 V short circuit**

- Todays electrical system with an operation voltage of 14 V is a single voltage distribution system.

  Therefore the consideration of short circuits between voltage levels are

- necessary in a dual-voltage system
- for an interim time period (until a single 42 V system will be established
- complex

The probability of these faults might be low but effects are hazardous.

- Critical case:
  Short circuit between the power busses without connection to ground
- Solution Concepts
  - Separation
  - Shielding (with shield to ground)
  - Current and voltage detection on both voltage levels
    - **central**
    - **decentral**
    - **per function**

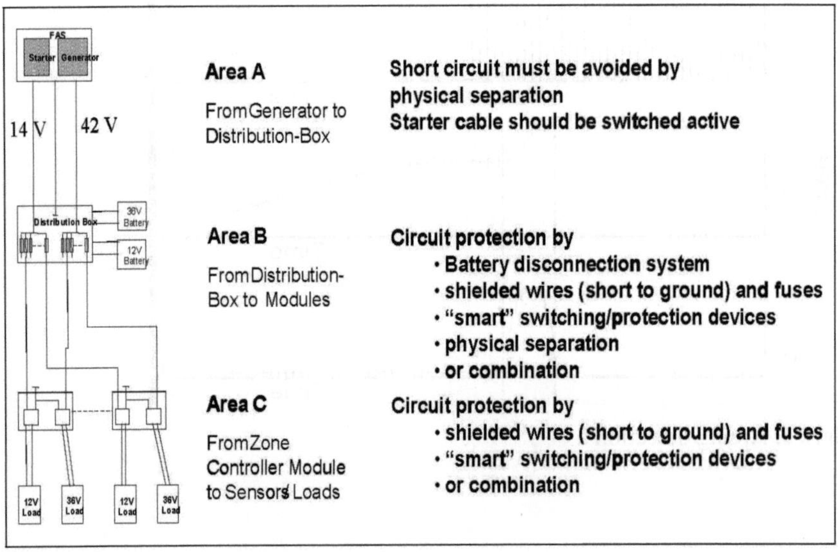

11 Short Circuit Protection Methods

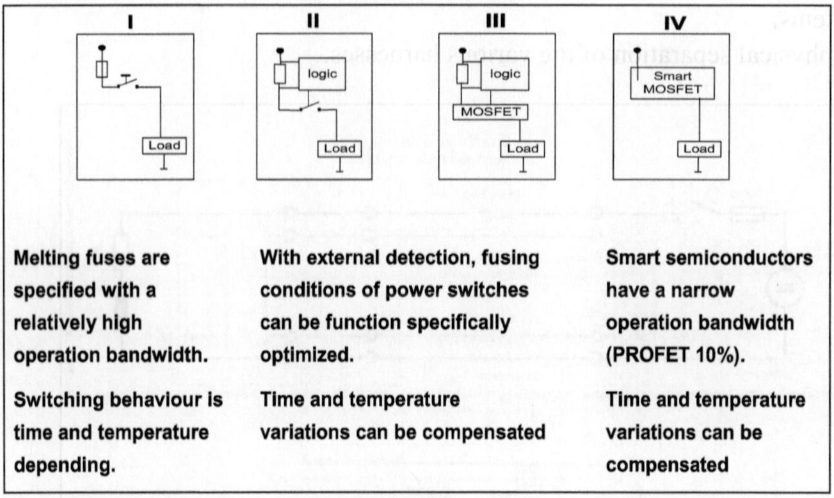

Fig. 14. Short Ciruit Protection Methods

12 Power Switching & Protection

Fusing of the two voltage levels against each other:

In general a short circuit to ground will be detected even with traditional blow fuses. This case doesn´t cause any problems. Blow fuses have a wide range of tolerance (operating range) – in case of leakage current there will be enough energy supplied to the failure, sufficiant to cause fire, before a blow fuse cuts off the circuit.

This case needs additional investigation. Our recommendation is to take advantage of electronic fuses, except in case of the need of galvanic separation.

The challenge to detect a short between the two power bus lines (14 to 42) needs additional electronic efforts.

Solution:

- current sensing on both – the supply line to the load as well as the line back to ground. In case of failure current to ground or to the other voltage level both currents are different and can be detected by the system. (Failure current detection);
- in order to detect serial arcing in connection systems or in case of line defects we recommend voltage sensing at the output lines at the central

electronic (supply) and at the same time at the load or device. This solution indicates smart loads or devices or at least smart connection systems;

- physical separation of the various harnesses.

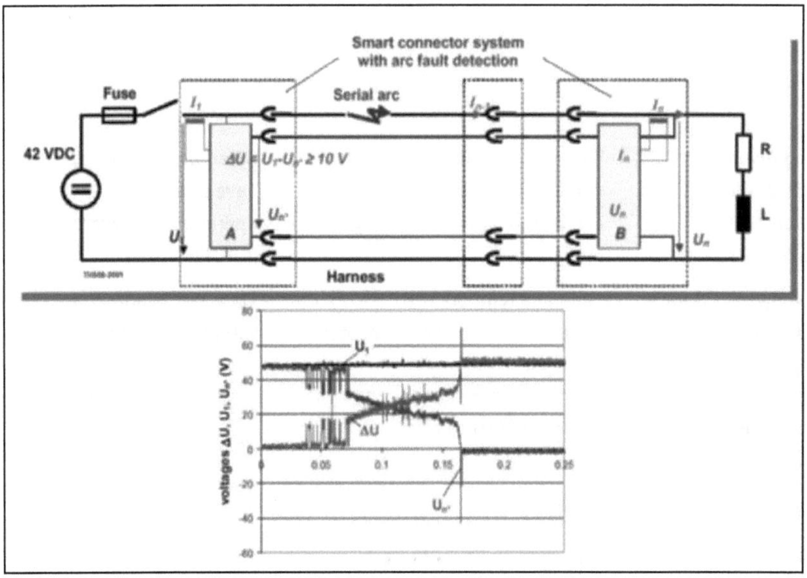

Fig. 15. Arcs @ 42V

13 How can the cost impact of System Protection be limited?

- Currently applied relays cannot be used for 42V-PowerNet
  different designs are required due to arcing
- Power switching and circuit protection by application of:
  42 V Relays plus fuses plus additional current detection
  MOSFET´s plus central protection
  SMART MOSFET´s
  MOSFET´s plus function related current and voltage detection

➔ Additional cost is just for detection electronics, which might be shared with devices for power management

- Overview

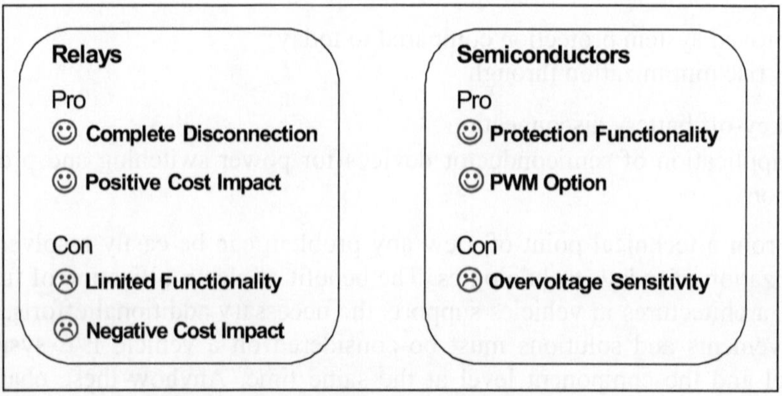

Fig. 16. Power switching and protection: pros and contras

14 Working Party Short Circuit

In September 2000 the Working Party Short Circuit (WPSC) was established by members of the Forum Bordnetzarchitektur in order to deeper discuss the issue of circuit/system protection.

It was decided to create a "guideline" for system protection.

Intention of the WPSC:

- prevent destruction and damage of the vehicle electrical system
- system protection should be at least as good as today
- only single faults will be considered
- it is not intended to create a standard or to specify a certain technology
- this paper should help OEMs and suppliers to decide on system protection measures

15 System Protection Summary

100 % System Protection will not be economically feasible.

Without additional detection devices, conventional fuses and devices with similar characteristics like circuit breakers or PTC's cannot perform system protection in case of an "inter-power-bus-short circuit" or arcing.

Strategy

Improved system protection compared to today
plus risc minimization through

- key-off battery disconnection
- application of semiconductor devices for power switching and protection

From a technical point of view any problem can be easily resolved by utilization of todays technologies. The benefit of electronification of future E/E architectures in vehicles supports the necessary additional efforts. Improvements and solutions must be considered on a vehicle E/E systems level and the component level at the same time. Anyhow these changes will cause additional cost added to the involved products.

For the implementation we predict a careful step by step introduction of 42 V loads in the case, where the 12 V system is at least overloaded.

Requirements for Introduction of the 42V-PowerNet

Volker Graf
Intedis GmbH & Co. KG

Abstract

This abstract will give an insight about the implementation of the 42V-PowerNet with emphasis on the impact and the prevention of electrical overloads or cable fires on wiring safety.

After we look at the conventional way of fusing the wiring harness in cars with thermal fuses and semiconductor drivers, we will show the upcoming challenges with the introduction of the dual voltage power network. Here we concentrate on fusing issues coming up in a dual voltage network since this seems to be the first step towards bringing 42 V into the motor car, since it will be impossible to adapt all necessary components at the same time to the higher voltage.

In connection we will show the risks arising through arcs, which can occur in DC networks with higher voltage. Some theory behind the arc and possible ways of dealing with it will be shown.

The mentioned factors will point out the requirements needed to fulfill the functions of the future PowerNet.

1 Introduction

Times like these, when car wiring systems are already burdened with EMC or the provision of a smoothly installed ground connection in regards to e.g. safety systems, the implementation of higher voltage would arise more challenges, which is partly shown in this document.

2 Optimising wire fusing

2.1 Demands on wire fusing

Main task of the fuse:

- A fuse has to protect the isolation of the wire from excessive temperature and hence from destruction/fire

Functions:

- A fuse has to conduct the normal current during use

In case of short circuit:

- Recognising the event of a short circuit/overload
- Switching off the short circuit current/separation of the overload.
- Safe handling of inductive energies ($E= 0{,}5I^2L$/ appears as arc)

2.2 Work Flow of fusing a wire

This section will basically show the procedure of fusing. Fig. 1. shows the steps to be taken to fuse a wire.

In order to estimate the requirements for cross sectional area (CSA) an engineer will do the following steps:

- Gaining knowledge of the current magnitude and its distribution against time (a larger CSA will be less sensitive against short high current pulses because of the higher heat capacity)
- Calculating the voltage drop in the cable along its length taking into account things like contact resistance, the number of contacts, etc.
- Determining the minimal CSA
- Determining the allowable temperature rise for the wire in question due to the environmental conditions (e.g. near the engine or inside the car; mechanical strain)
- CSA still sufficient for current and the mechanical strain ?
- Using look up tables the engineer will choose a suitable fuse
- Check on the fuse blow with the wire connected depending on the load dynamics, the temperature of the fuse environment and fuse type

In many cases there are several wires connected parallel to one fuse (due to cost saving pressures, insufficient fuse sockets available, etc.), therefore the following questions have to be answered:

- is there more than one consumer on the fuse ?
  has the fuse to be upgraded (due to overall currents) ?
  If the fuse is upgraded, then the CSA has to be determined again and if necessary upgraded as well. **Should this be impossible, then the connection of several consumers to one fuse is not allowed !**

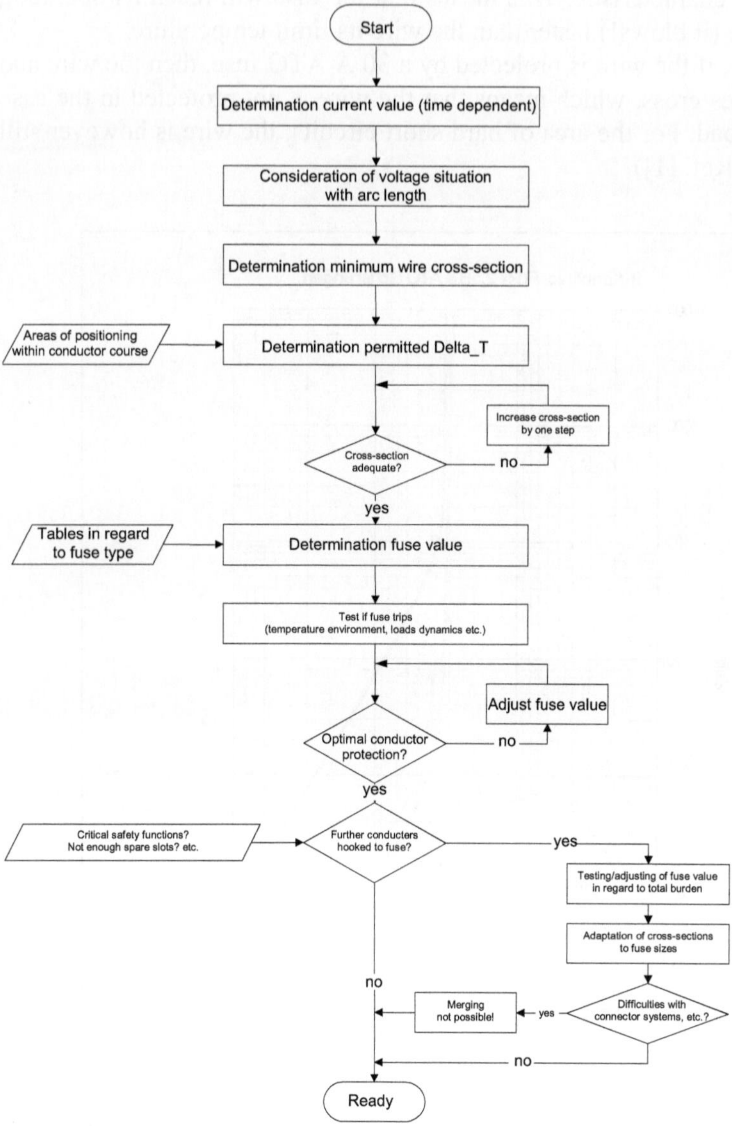

Fig. 1. Flow diagram for fusing a wire

2.3 Example: Wire and conventional thermal fuse

With the following example we show a FLRY 2,5 mm² (Type B) in a fusing scenario. As can be seen for a wire temperature rise of 40 K there is a blue curve, and if a 25 A ATO fuse (Littelfuse) is chosen (red line). The wire is protected against short circuit as the fuse characteristic lies to the left of wire characteristic. This means that the fuse will reach its operating temperature (it blows!) faster than the wire its limit temperature.

However, if the wire is protected by a 30 A ATO fuse, then the wire and the fuse lines cross, which means that the wire is not protected in the case of an overload. For the area of hard short circuits, the wire is however still protected. (Ref. [1])

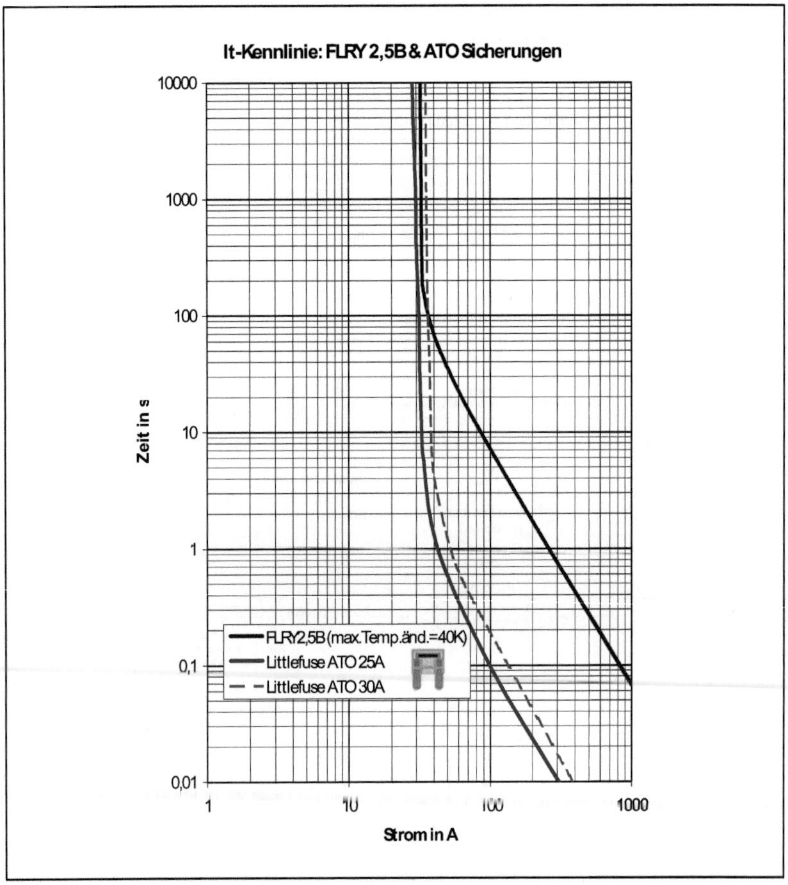

Fig. 2. Example of a wire/fuse combination

2.4 Example: Wire and High Side Driver

Modern semiconductor switches are commonly equipped with MOSFET power stages and include self protection features such as over temperature cut out and over current limiters. Hence they can favourably be used to combine the function of the relay and the wire fuse in one component.

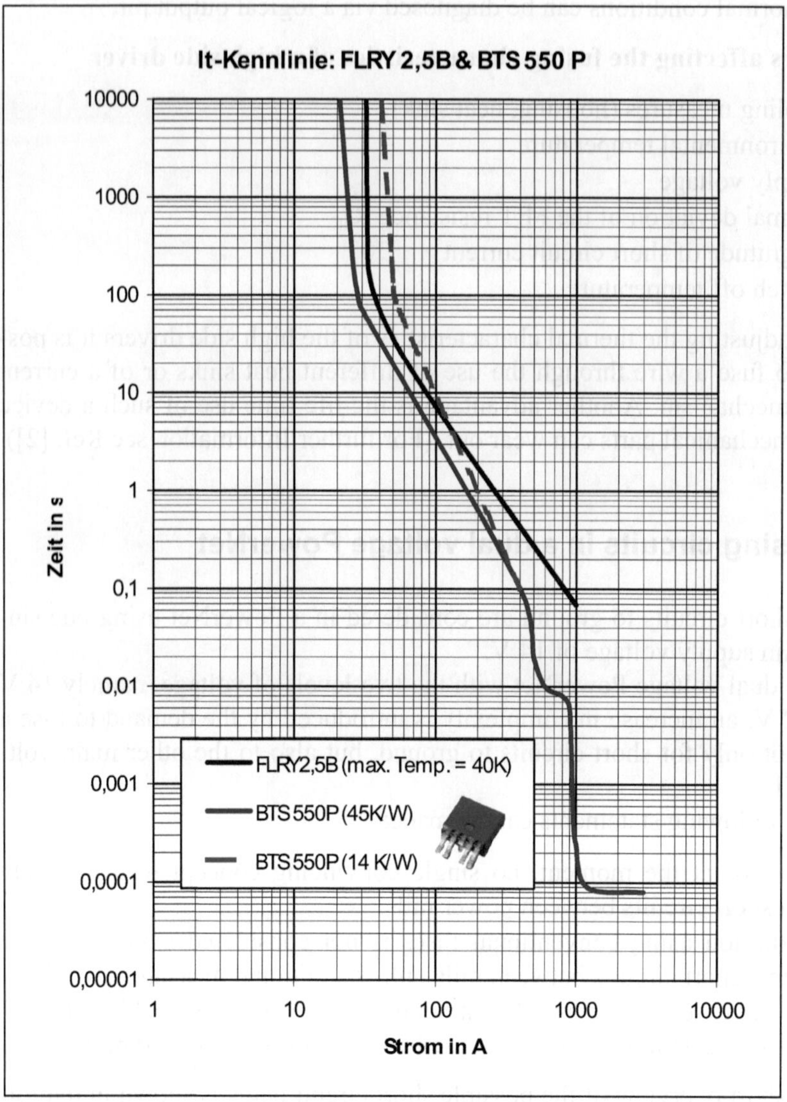

Fig. 3. Example of a wire/high side driver combination

Important characteristics of semiconductor drivers

- Internal over temperature cut out
- Built in over current limitation
- Short circuit recognition
- Several protection mechanisms against over/under voltage, short circuit and overload, prevent the destruction of the semiconductor
  Abnormal conditions can be diagnosed via a logical output pin

Factors affecting the fusing characteristics of a high side driver

- Cooling measures (housing, heat sink)
- Environmental temperature
- Supply voltage
- Normal deviation of the FET resistance R_{on}
- Magnitude of short circuit current
- Switch off temperature

By adjusting the thermal characteristics of the high side drivers it is possible to fuse a wire through the use of different heat sinks or of a current sense mechanism. Another advantage is the life time use of such a device as no mechanical parts can wear out. (For further Information see Ref. [2])

3 Fusing circuits in a dual voltage PowerNet

Only short circuits to ground are considered in a PowerNet using one single main supply voltage of 14 V.

In a dual voltage PowerNet with the two levels of voltage, namely 14 V and 42 V, an increase in complexity is introduced by the demand to fuse a wire not only for short circuits to ground, but also to the other main voltage rail.

The following statements can be made:

- There is, at the moment, no single convincing concept to deal safely with short circuits between power rails.
- A solution using conventional fuses is not considered a safe measure against short circuits between rails unless additional measures are taken
- The increased use of intelligent semiconductor drivers enables alternatives in the area of fusing to be used which will be explained briefly.

An short overview of the possible short circuit paths is shown in the following Fig. 4.:

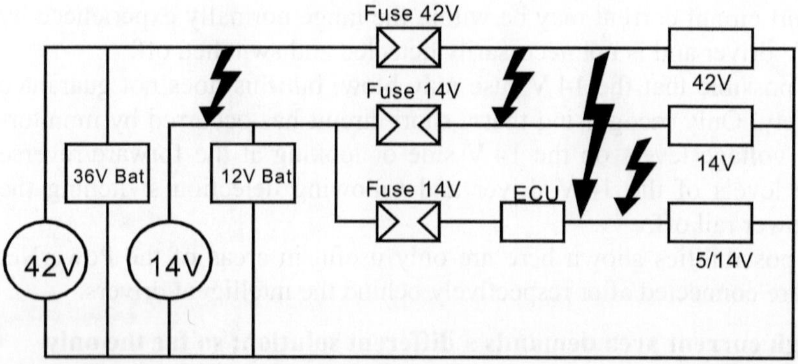

Fig. 4. Short Overview of possible short circuits between 14 V and 42 V rails

3.1 Dual voltage net fusing using intelligent semiconductor drivers

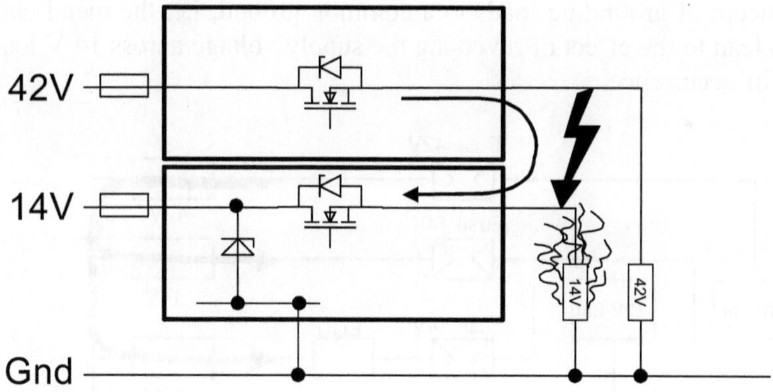

Fig. 5. Short circuit of power rails using semiconductor drivers

The following scenarios can be thought of:

Low impedance short circuit

High over current from the 42 V to the 14 V side is recognised by the 42 V driver, which switches the power off. Damage to the 14 V side is likely to be non-existent as fast switch off times can be realised.

High impedance short circuit

The short circuit current may be within the range normally experienced by the 42 V driver and is not necessarily detected and switched off.

It is possible that the 14 V fuse may blow, but this does not guarantee any safety. Only recognising that a short circuit has occurred by monitoring the voltage levels on the 14 V side or looking at the forward/reverse voltage levels of the 14 V driver and following detection switching the 42 V power rail off.

The possibilities shown here are only useful in areas of the PowerNet which are connected after respectively behind the intelligent drivers.

The high current area demands a different solution; so far the only practical way of dealing with this problem seemed to be a physical separation of the two power rails in order to avoid short circuits being created in the first place.

3.2 Loss of common ground

The concept of grounding loads to a common ground, i.e. the metal chassis, can lead to the effect of reversing the supply voltage across 14 V loads in case of occurrence.

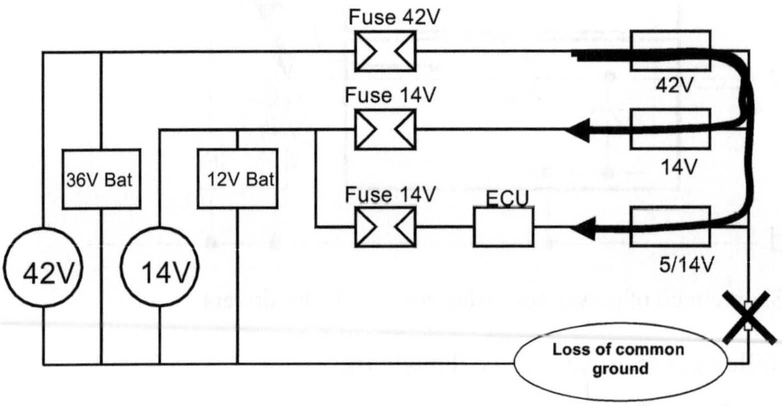

Fig. 6. Reversing the supply voltage in case of loss of common ground

3.2.1 Ground concept to avoid voltage reversal

Voltage reversal can be avoided by
Using separate ground paths for 14 V and 42 V grounds (if necessary separate bolts)

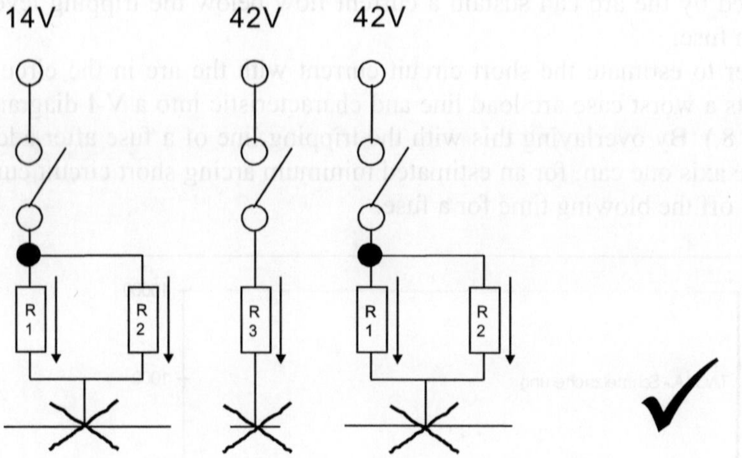

Fig. 7. Avoiding voltage reversal at the loads in dual voltage power rails

4 Arcs in 42V PowerNets

With the introduction of the higher voltage supply rail there is an increased likelihood of arc tracking occurring. Arcs can be observed when current of a sufficient magnitude is interrupted as when

- Separating connectors under power
- Changing a fuse under power

(normal separation of switching contacts, such as relays, is not further considered here); but also through the separation of current carrying wires under fault condition such as

- Separation of leads during an accident
- Bad contact points

The high energy density of he arc is dangerous because of its high heat output and able to cause damage to even start a fire.

4.1 Theoretical Background

4.1.1 Arcing under short circuit conditions

Arcing under short circuit conditions poses the problem that the impedance represented by the arc can sustain a current flow below the tripping level of a given fuse.

In order to estimate the short circuit current with the arc in the circuit one inserts a worst case arc load line and characteristic into a V-I diagram (See Fig. 8.). By overlaying this with the tripping line of a fuse after adding a time axis one can, for an estimated minimum arcing short circuit current, read off the blowing time for a fuse.

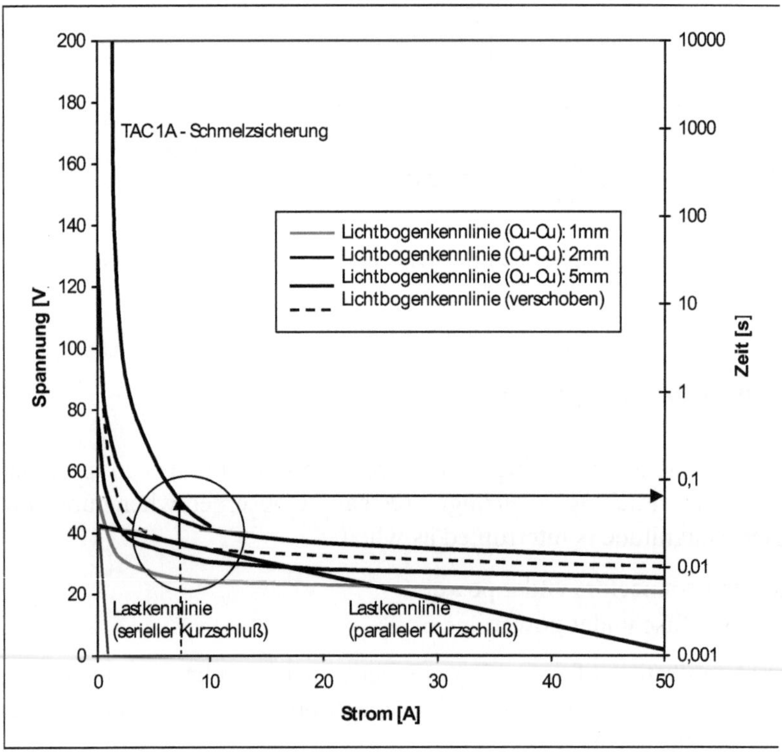

Fig. 8. Example for estimating a short circuit current and fuse-blow time under arc conditions

When using higher current fuses there is the risk that the arc is an impedance of a sufficient magnitude that efficient blowing currents can not be guaranteed.

For further Information see Ref. [3]

4.1.2 The arc „in circuit"

The arc which occurs in series with a current limiting load is the case mainly experienced in real life. It is characterised by

- A current which is lower than the normal load current
- An arcing voltage in the range from 16V to 42V (in a 42V net)
- An arc power which is the product of current times the arcing voltage

The maximum arcing power to be expected can be calculated by

$$P = U * I = \frac{U_0}{2} * \frac{I_0}{2} = \frac{U_0 * I_0}{4}$$

$$U_0 = SupplyVoltage$$

$$I_0 = LoadCurrent$$

The next Figure shows the arc power in series with a consumer

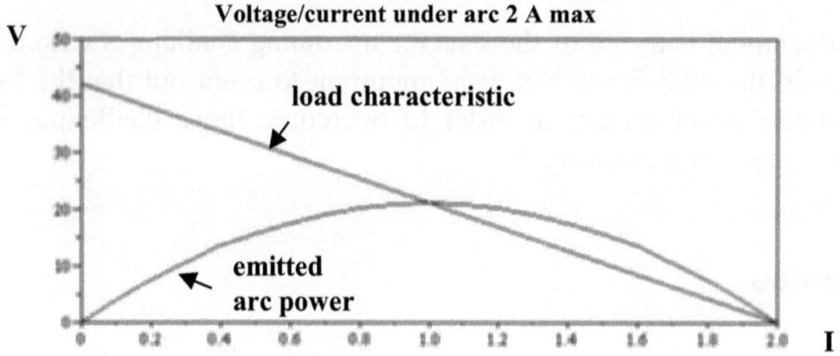

Fig. 9. Example of a 2 A load line and the corresponding arc power (Ref [1])

The here shown cognitions regarding the arc phenomena are used for the theoretical foundation to create a arc strategy for future PowerNet developments.

4.2 Dealing with arcs

1. Recognising /Localisation of the arc
2. Suppressing the arc

4.2.1 Examples for ongoing developments of arc recognition

- Voltage supervision in control units to determine drops pointing to the concurrency of arcs
- Monitoring a isolated wire inside main supply cable
- Monitoring the current to a load for the characteristic drop
- Monitoring the currents in parallel supply lines (resulting to a short circuit of the possible arc > Prevention) (Ref [4])
- Monitoring for noise spectrum of the arc

4.2.2 Extinguishing the arc

The Following acknowledged measures have to be taken to extinguish the arc, and the most effective means to do this, is to stop the current flow.

- Switching of the current controlling switch in the circuit
- Switch off/Switch on (~PWM)
- Crow- Baring a fuse in the circuit

5 Forecast

Since this document is trying to show us the upcoming challenges with the introduction of the 42V-PowerNet, it is important to point out that the future automobile development, in order to overcome those challenges is most effective by using electronics.

6 References

[1] Weidenbacher C, Absicherung mit und ohne Schmelzsicherungen- Grundlegende Techniken, Oktober 2001, Intedis GmbH & Co. KG
[2] Graf A, et al., Sicherungsersatz mit PROFET Highside Leistungsschaltern, Siemens AG
[3] Hetzmannseder E, et al., 42 VDC Arcing Faults – Physics & Test Methods, April 2001, EATON Corporation
[4] LEONI Patent [101 49 458.0-34]

Test Methods for Fusing Devices

Ronald Zörn, Jürgen Scheele, Werner Hinrichs

PUDENZ GmbH, 27243 Dünsen Germany

Abstract

The most used fuses for automobiles are defined in specifications. These specifications or standards describe the fuses, the requirements and the tests for the fuses. But the test methods are not exactly defined. So it is possible to use test equipment which affect the tests in a bad way. This is specific important for tests in the new 42V-PowerNet. The high test-voltage of 58V favour to form a arcing at the separation of the current. This arcing must be controlled and extinguished by the fuse-link. This will be support by “Ohm´s law” and additional by the fuse-link. So it is very important, that the test device don´t work against this behaviour.

1 Introduction

Fuses for automotive applications must be tested in different ways. These tests are defined for example in the DIN 72581-3 or ISO 8820-5. The head-tests are the environmental-tests, the chemical-resistance-tests and the electrical-tests. The environmental-tests are necessary for the behaviour of ageing. Beside the environment-tests the chemical tests check the resistance of the fuse-link for the most existing liquids of a car. A contact with this liquids can affect the fuse in their function. But the most important test for the fuse is the electrical-test

Why?

The main duty of the fuse is the protection of the electrical circuit. The fuse must protect the circuit in the specified limits of the specification. The other requirements are side conditions. All tests finished with the electrical test which give the final statement about the fuse.

2 Principle for the release of a fuse

Two reasons are responsible for the release of a fuse. The current and also the temperature. The temperature divide in the ambient-temperature and the process-temperature dependent on the current. Both together leads to the melting of the fuse.

The current heats up the melting element in the fuse. With the increasing temperature the resistance of the melting element rise up, and the current heats the melting element higher. In a short or long time, dependent on the current, the melting element melts and interrupt the circuit. This behaviour is not dependent on the nominal voltage in the circuit. The voltage drop of the fuse only dependents on the fuse-resistance and the current.

$$U_{fuse} = R_{fuse} * I_{circuit} \quad (1.)$$

A very high current, like a short-circuit changes the melting-element characteristics. The melting element will not warm up, instead of this the current evaporate the melting element and "blast" the material.

Table 1. Voltage- and current-limits for arcs

contact-material	U_{min} [V]	I_{min} [A]
Silver	12	0,4
Gold	15	0,3
Palladium	15 - 16	0,8 - 0,9
Platinum	17	0,8 - 1,0
Copper	12 - 13	0,4
Nickel	14	0,4 - 0,5
Wolfram	15 - 16	0,8 - 1,2
Carbon	20	0,01 - 0,02

Both processes are independent from the voltage in the circuit. The voltage about the fuse is very low and only in the moment, where the fuse interrupts the current, the voltage rises up to the value of the circuit. If the voltage level is high enough an arc can keep alive. Now the fuse must switch off the arc to protect the circuit. This is in the new 42V-PowerNet harder then in a 32 V powernet, because the voltage-drop about the circuit is higher and the limits to the minimum voltage of an arc is smaller. The minimum voltage of different materials are shown in Table 1.

2.1 Tests for the 42V-PowerNet

The new 42V-PowerNet opens the door to many new applications and systems in the car. The output in the cars rise up that the limits of the 12 V power-network are reached. So it was logical to increase the voltage to a higher level. But the higher voltage brings also the possibility of arcs.

In the moment of release the fuse-link must be able to switch off the arc, without an defect. The only way for the fuse is to take away the energy of the arc. The arc is depended on the voltage and the current of the circuit.

But the problems with an arc are not limited to the fuse-links. All components like cables, connectors and terminals can have this problem.

The new draft version of the specification DIN72581 takes into consideration the new higher voltage and changed the requirements. The nominal voltage is now 42 V and the max. test-voltage is 58 V. But the requirement are the same as for 32 V.

All tests and requirements for the fuse based on three electrical behaviours. The over-current test , the short-circuit test and the load-test. The difference between the tests is the altitude of the current.

By an over-current test the current is controlled and depended on the nominal current of the fuse. This tests defined the time-current-characteristic of the fuse. The maximum test-current is 6 or 10 times the nominal current.

The short-circuit-test is defined by a prospective current. This current is the theoretical current, which will flow if a copper-bridge will be used instead of a fuse. The real current about the fuse is lower than the prospective current. This is depended on the higher resistance of the fuse to the counterpart of the copper-bridge. The test current will only be controlled by the resistance of the circuit, the fuse and the source of car-batteries. The electrical circuit follows the “Ohm´s law”.

A load-test is a life-time-test. The different tests will often be combined with other factors like temperature, load-cycles or other requirements.

With more and more new applications and devices in the cars the requirements changed. The new devices are electronic adjust and change the internal resistance within limited ranges. So we have a variable resistance in the circuit, which can favour the arc.

2.1.1 Effects from test-equipment

A statement in the DIN 72851-3 read: “The voltage and current must be constant at a range of ±1%, if no other requirements are specified”. This comment can be misunderstand. If we use an electronic-load in the mode of operation “constant-current” for example, we can affecting the test in a bad way.

The problem with an electronic-load in this mode is, that the load changed the internal resistance to hold the current constant. In this case, the load changed the resistance in a limit from 10% to 90% in 0,5 ms, but the arcing-time will be near by 1 ms. So the load is faster than the arcing-time and supports the life of the arcing. This means that the load works against the “Ohm´s law”. A “burning” arc has a higher resistance as the melting element of the fuse. When the resistance increases, the current of the circuit must be limited. But the electronic-load reduces her internal resistance faster than the resistance of the arc can rise up. With this behaviour the arc will be hold on life, because the current can not reduced. The electronic-load controlled the current against the fuse. So the electronic-load supports the arc and increases the power in the fuse. This power is so high, that it will be impossible for the fuse to put out the arc. The fuse will be melting or burning.

The following diagram show the different of the cutoff-power between two modes of operation of the electronic-load.

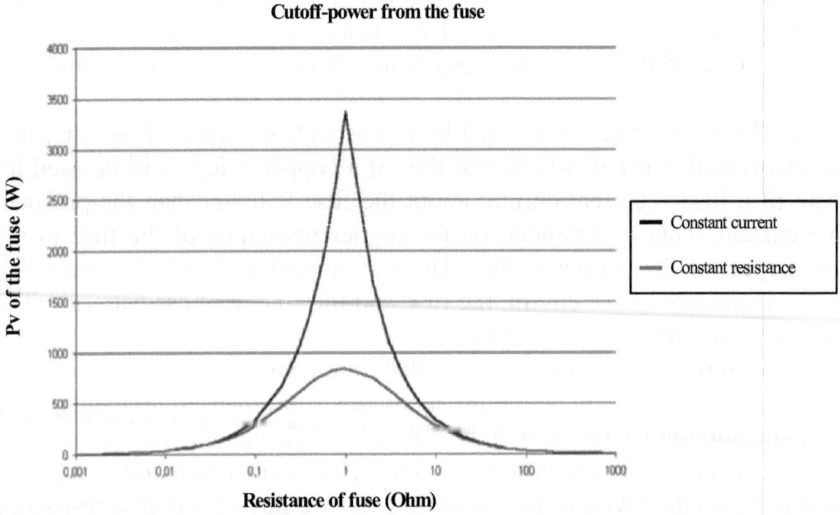

Fig. 1. Cutoff-power from the fuse

The difference of the modes are the current-limitation. The "constant-current" mode adjust the current always to the selected value. The "constant-resistance" mode adjusts a fixed resistance. When the resistance is fixed, the current can be limited by the higher resistance of an "burning" arc. With the limited current the fuse can put off the arcing and interrupts the circuit.

The effect of the "constant-current" mode is higher, if the test-current for the fuse is smaller. This is depended on the release of a fuse. A low over-current leads the fuse to warm up and a smaller separation comes about the melting-element. So it is easier to ignite an arc. Additional the warmed up material of the fuse can not cool down the arc.

3 Summary

The test with an electronic-load in "constant-current" mode is not a realistic test, because the acting-time of the device is faster than the arcing-time. The only device which has approximately the behaviour on an electronic-load in "constant-current" mode is a DC-DC-Converter for the new 42V-PowerNet. These devices can limit or rise up the current in limited control- and time-ranges. But the time-ranges for a DC-DC-Converter are 2ms and not 0,5ms like the electronic-load. Thus the acting-time is lower than the arcing-time. Also the voltage control range of a DC-DC-Converter is limited, that it cannot work at a voltage lower than 10 V.

The statement in the DIN 72581-3 means not to use an electronic-load which hinder the fuse to interrupt the circuit. The workshop for standardiziation works on a new interpretation and wording of this statement and also on other parts of the specification.

Also we must define, which requirements are realistic in the car:

- Is it possible, that a device has a behaviour like the electronic-load ?
- Can the 42V-PowerNet deliver the power for this effect or will the voltage break down ?

This points must be discussed and defined. Then it will be possible to determine tests and requirements for the fuses.

MIX
Papier aus verantwortungsvollen Quellen
Paper from responsible sources
FSC® C105338

If you have any concerns about our products,
you can contact us on
ProductSafety@springernature.com

In case Publisher is established outside the EU,
the EU authorized representative is:
Springer Nature Customer Service Center GmbH
Europaplatz 3, 69115 Heidelberg, Germany

Printed by Libri Plureos GmbH
in Hamburg, Germany